Also from *New Scientist*

How to Fossilize Your Hamster: And Other Amazing Experiments for the Armchair Scientist

Why Don't Penguins' Feet Freeze? And 114 Other Questions

Does Anything Eat Wasps? And 101 Other Unsettling, Witty Answers to Questions You Never Thought You Wanted to Ask

Stroke a Martian: And 99 Other Things to Do Before You Die

100 Things to Do Before You Die (Plus a Few to Do Afterwards)

DO POLAR BEARS GET LONELY?

DO POLAR BEARS GET LONELY?

And Answers to 100 Other Weird and Wacky Questions about How the World Works

NewScientist

Edited by Mick O'Hare

A HOLT PAPERBACK

Henry Holt and Company New York

Holt Paperbacks
Henry Holt and Company, LLC
Publishers since 1866
175 Fifth Avenue
New York, New York 10010
www.henryholt.com

A Holt Paperback® and ® are registered trademarks of
Henry Holt and Company, LLC.

Library of Congress Cataloging-in-Publication Data

Do polar bears get lonely? : and answers to 100 other weird and wacky questions about how the world works / New Scientist edited by Mick O'Hare.
p. cm.
Includes bibliographical references and index.
ISBN-13: 978-0-8050-8988-2
ISBN-10: 0-8050-8988-8
1. Science—Miscellanea. I. New Scientist. II. O'Hare, Mick, 1964–
Q173.D623 2009
500—dc22 2008041000

Henry Holt books are available for special promotions and premiums. For details contact: Director, Special Markets.

Originally published in Great Britain in 2008
by Profile Books Ltd., London

First Holt Paperbacks Edition 2009

Margin illustrations by Brett Ryder
Designed by Kelly S. Too

Printed in the United States of America
1 3 5 7 9 10 8 6 4 2

CONTENTS

INTRODUCTION

It's a fair bet that you've never considered what compels you to choose random numbers in the Lotto draw. But now that we've told you that you do, you'll almost certainly want to know why. If you are truly perplexed, turn now to page 212. And has it ever struck you that if you go back forty generations your ancestors will total more than the number of people who have ever lived? How can that be? Find out the flaw in the logic on page 222. And what about eating bits of yourself in order to stay alive . . . you might not like what we suggest on page 48. Come to think of it, *we* don't like what we suggest.

For nearly fifteen years the readers of *New Scientist* magazine have been contributing their astounding knowledge to the "Last Word" column of science questions and answers. We now know why cheese goes stringy and what time it is (or isn't) at the North Pole. We know how to weigh our heads and we certainly know why penguins' feet don't freeze. We've also put to rest a couple of urban myths along the way—see page 59 to find out if human hair and fingernails really do continue to grow after death. And we've also been happy to admit our own errors. See page 38 to find out just how humiliated we were by a few broken drinking glasses.

But that's all part and parcel of scientific investigation. You propose a hypothesis, bash it about a bit, run a few experiments on it and then reject it or accept it while fine-tuning it along the way. And like all great scientists, that is exactly what the contributors to this book have done—deduced answers from the evidence available and then been supported or contradicted by their peers, which is what makes the "Last Word" column—and this book—such fascinating reading.

Both in the weekly magazine and online, *New Scientist*'s community of "Last Word" readers continues to come up with the answers to the world's strangest questions. And more people are always welcome. You can pose your own questions or answer new ones (or even contradict those who have gone before) by buying the weekly magazine or visiting the website (http://www.newscientist.com/lastword). There you can join in the forum, read the blog, or simply offer your knowledge in answering some of life's astounding conundrums. You may even become the star of the next book!

In the meantime enjoy this one, in which we seem unduly concerned by thirst—you can find out in chapter 5 whether fish, sharks, or spiders get thirsty and, along the way, satisfy your own thirst for the world of scientific trivia.

MICK O'HARE

A big thank-you is due to Jeremy Webb, Lucy Middleton, Ivan Semeniuk, the production, subbing, art, Web, press, and marketing teams of *New Scientist*, James Kingsland, Frazer Hudson, and Paul Forty and Andrew Franklin among many people at Profile Books for their tireless efforts in the creation of this volume. I also thank Robin Dennis, my editor at Henry Holt and Company in New York, for bringing this project to the States. Thanks as well to the team at Holt, including Justin Golenbock, Emi Ikkanda, Tom Nau, Rita Quintas, and Kelly Too, for their hard work. Special thanks are also offered to Sally and Thomas for their patience while this book and its predecessors were created. Finally, good luck to Ben Usher on his travels.

DO POLAR BEARS GET LONELY?

1 FOOD AND DRINK

TWIN CHICKS

Upon cracking open my breakfast boiled egg, I found a whole new egg inside. It was not a double-yolked egg, it was a double-egged egg—a completely new egg with a shell and yolk inside another. Can anybody explain it?

Liam Spencer

An egg within an egg is a very unusual occurrence. Normally, the production of a bird's egg starts with the release of the ovum from the ovary. It then travels down the oviduct, being wrapped in yolk, then albumen, then membranes, before it is finally encased in the shell and laid.

Occasionally an egg travels back up the oviduct, meets another egg traveling down it, and then becomes encased inside the second egg during the shell-adding process, thus creating an egg within an egg. Nobody knows for sure what causes the first egg to turn back, although one theory is that a sudden shock could be responsible. Eggs within eggs have been reported in hens, guinea fowl, ducks, and even Coturnix quail.

Incidentally, it is especially unusual to encounter this

phenomenon in a shop-bought egg, because these are routinely candled (a bright light is held up to them to examine the contents), and any irregularities are normally rejected.

Alex Williams

As the curator of the British Natural History Museum egg collection, I've come across quite a few examples of egg oddities. Double eggs (as opposed to multiple-yolked eggs) are less common than some other oological anomalies and consequently the "ovum in ovo," as the phenomenon described here is known, has attracted specific scholarly attention for hundreds of years.

The Dominican friar and polymath Albertus Magnus mentioned an "egg with two shells" as far back as 1250 in his book *De animalibus*, and by the late seventeenth century pioneering anatomists like William Harvey, Claude Perrault, and Johann Sigismund Elsholtz had also given the phenomenon their attention.

Four general types occur—variations of yolkless and complete eggs—but this form in which a complete egg is found within a complete egg is relatively rare. Several theories have been proposed for the origin of these double eggs, but the most likely suggests that the normal rhythmic muscular action, or peristalsis, that moves a developing egg down the oviduct malfunctions in some way.

A series of abnormal contractions could force a complete or semi-complete egg back up the oviduct, and should this egg meet another developing egg traveling normally down the oviduct, the latter can engulf the former; more simply, another layer of albumen and shell can form around the original egg.

Often when no yolk is found within the "dwarf" or interior

egg, a foreign object is found in its center. This object has served as a nucleus around which the albumen and shell were laid down, in a process not dissimilar to the creation of a pearl.

Anybody interested in learning more about this subject should try to find a copy of *The Avian Egg* by Alexis Romanoff and Anastasia Romanoff (New York: John Wiley & Sons, 1949) and read pages 286–95.

Douglas Russell

Curator, Bird Group, Department of Zoology

The Natural History Museum, Hertfordshire, United Kingdom

A ROUND FIGURE

Why do bottle caps on beer bottles—at least the few hundred thousand that I have drunk from—always have twenty-one sharp bits?

Volker Sommer

We have three explanations for this one. We're still waiting for a bottle-top aficionado (of which there seem to be many) to rule between them.—Ed.

The bottle cap on any bottle is regulated by the internationally accepted German standard DIN 6099, ensuring all bottle caps are the same. Along with specifying the diameter of the bottle neck, the form of the rim around which the cap is crimped, and the materials the cap may be constructed from, this document specifies the form of the crimp. One requirement is that the closure be sufficiently circular to maintain a tight seal all around the circumference, which implies a high

number of crimps (and thus points). It must also be robust, however, which implies reducing the number of crimps to give each crimp a larger bearing surface. Using twenty-one crimps is a good compromise between these requirements and is mandated in the standard. As to why it is twenty-one crimps rather than twenty or twenty-two, the best answer is simply "because it is."

S. Humphreys

Through trial and error, William Painter, the inventor of the crown cork, or bottle cap, discovered that the optimum number of teeth on a mold made of steel for securing carbonated drinks was twenty-four. He registered a patent for his design and for many years the twenty-four-tooth capping mold was standard. However, around 1930 the steel mold came under threat from a cheaper version made of tinplate. This newer mold could not win a patent if it also had twenty-four teeth, so it was changed to twenty-one to avoid infringing the original design. The new figure is the smallest number of teeth needed to prevent leaks and is now used across the world.

Chitran Duraisamy

The crown cap was patented by Painter on February 2, 1892 (U.S. patent 468,258). It originally had twenty-four teeth and a cork seal with a paper backing to stop drink and metal touching. The current version has twenty-one teeth.

The twenty-four-tooth caps were originally fitted to bottles one by one using a foot-operated press. When automatic machines were adopted, the crown caps were loaded into circular feed tubes and the twenty-four-tooth caps frequently jammed. With an uneven number of teeth this doesn't happen, and because the sealing quality of twenty-three teeth was no better than twenty-one, the smaller number was adopted.

The height of the crown cap was also reduced and specified in the German standard DIN 6099 in the 1960s. This also defined the "twist-off" bottle cap that is widely used in the United States.

Barry Painter

CEREAL KILLER

Most healthy people I know eat cereal or fruit for breakfast. This gives complex carbohydrates for long-term energy. But I have a physical job as a gardener and I know if I rely on this intake I'll be ravenous by 10 A.M. On the other hand, if I have eggs, I'll be fine until midday. Clearly I need protein, but that shouldn't give me energy. What is going on, and is this common?

Steve Law

It may be that your hunter-gatherer ancestry is responsible for the favorable response to your morning serving of eggs. In the course of human evolution we have become physiologically adapted to the diet that prevailed for most of that time: that of a hunter-gatherer. This diet is assumed to have been dominated by lean meats, fruits, and vegetables. Cereal grains, on the other hand, are a relatively new addition to our diet, having found their place on the dinner table with the onset of the agricultural revolution only ten thousand years ago.

It has been suggested that our pre-agricultural diet is the best way to support healthy physiological function, including improved energy production and appetite control. One of the characteristics of this diet is a low "glycemic load,"

which means glucose is released slowly into the blood as food is digested. Another is a higher level of lean protein than that eaten by modern humans. These characteristics are found in your eggs, whereas most breakfast cereals and fruit have higher glycemic loads and lower protein content.

The low glycemic load of your meal may help to stabilize your blood sugar level, sharp drops of which precede an increase in appetite. The protein in eggs is also a strong inducer of cholecystokinin, a gut-derived satiating hormone. And carbohydrate is not the only source of energy in our diets. The fat in your breakfast eggs provides approximately double the energy of carbohydrate, albeit in a slow-release form.

Benjamin Brown
Technical Research Officer
Health World, Queensland, Australia

SACRED DNA

Animals and plants share a common genetic ancestry, so perhaps vegetarians who refuse to eat meat on ethical grounds should avoid anything that has DNA at all. Is this feasible? Could anybody suggest a menu?
Richard Ward

I'm not aware of any living organisms that don't have DNA, so you'd have a hard time eating any tissues or cell cultures. You could try eating RNA viruses, but you'd need to produce them in a cell culture, which generally requires animal serum to keep the cells alive. Your food wouldn't contain DNA, but you would have used dead animals to produce it.

One cheat that springs to mind is red blood cells. In many species, including humans, the nucleus and mitochondria are removed from these cells during the maturation process. This is to make room for more hemoglobin, the iron-bound protein that carries oxygen. Because the nucleus and mitochondria contain all the cell's DNA, you could argue that provided you don't kill the animals, drinking their blood is the ultimate vegetarian diet. You'd need to filter out the white blood cells, which still have plenty of DNA, but the rest of the blood components would be fine. They'd provide you with protein, some sugars, and vitamins, but probably more iron than is healthy.

If that doesn't sound appealing, consider totally (bio) synthetic foods. Biologists routinely construct yeast and bacterial lines designed to churn out large quantities of a specific protein or other biological molecule. I assume it would be possible to scale this production up to produce sufficient quantities of purified proteins, sugars, and so on to act as a food source. Don't expect it to be tasty, though: the proteins and sugars produced would be purified from the culture as crystalline powders. I'm not sure whether it's possible to produce fats like this without killing the cells, but if you did the result would either be oil or a pretty nasty goo. Also, maintaining the cultures required to produce this stuff would rely on antibiotics to kill contaminant organisms, so going against the spirit of the idea.

Many, perhaps all, of the various vitamins and other nutrients we require could probably be synthesized in similar ways, given time and cash. The various mineral compounds we need—iron, copper, zinc, iodine, and so on—are probably available from a good synthetic chemist. And, of course, you could drink milk. It's a complex mixture of secreted

proteins, fats, sugars, and pretty much everything else you need to stay alive. It may contain cells from the animal which produced it, but you could probably centrifuge these out.

Christopher Binny

All I can come up with is a dish of baked retrovirus served on a water biscuit made from purified starch, fried in a purified fat of choice, and seasoned with salt and vinegar. For the sweet course you might try a sorbet of snow sweetened with a purified sugar, honey, or syrup, a touch of citric acid for bite, and with added vitamins, trace elements, and essential oils to taste. It should be washed down with any spirit, or any wine or beer filtered to remove yeast traces.

Bryn Glover

I found the following information on the wall of the Johnson Space Center in Houston, Texas. One cubic meter of lunar soil contains enough of the right elements to make a cheeseburger, an order of fries, and a fizzy drink. That would contain no DNA, but might be a little expensive.

Graham Kerr

I considered this some years ago and put my conclusions in the form of a cookery book, available online at http://www.cs.st-and.ac.uk/~norman/Shorts/inorganic.html.

Norman Paterson

To whet your appetite, here's a recipe from Norman Paterson's book.—Ed.

For four malachite burgers you will need:

> Four slices of Welsh slate
> 1 kilogram of malachite
>
> Cut the slates in two. Break up the malachite with a sledgehammer. Divide the malachite equally among four slates and cover with the remaining four. Bake at 2200°F for twelve hours, by which time the malachite should be a beautiful bubbly green. Cool and eat. Excellent for picnics, as they can be prepared the century before. A dry, gritty flavor.

Most people who are vegetarian on ethical grounds oppose killing animals. They are rejecting the senseless deaths of the animals and the inhumane way the animals are treated, rather than worrying about similar DNA. Vegetarians have nothing against eating vegetable matter and fungi because these have no central nervous system and thus cannot experience pain.

Ceridwen Fitzpatrick

If all plants and animals have common DNA ancestry, then perhaps we are all vegetarians. Or if we are all also vegetables, maybe the world is awash with cannibalism.

Or perhaps vegetarians can eat their neighbors without feeling too much guilt. By the "common DNA" logic this is no more or less cannibalistic than eating a radish. The only solution to all these dilemmas would be for every creature to subsist purely on nonliving minerals and nutrients. Nonhuman animals, however, are unlikely to stop eating what they want.

Brian Falconer

STUFF THAT

I cooked some poultry stuffing and left it in a bowl in the fridge overnight, covered with aluminum foil. In the morning there were holes in the foil where it had touched the stuffing, which was stained black under each hole. Uncooked stuffing does not produce this effect, and it makes no difference whether the stuffing is cooked inside the bird or separately. What is going on here, and are the black stains poisonous?
Andrew Stiller

Without its submicroscopic insoluble skin of oxide, aluminum cookware would catch fire easily. Fortunately, this is not usually a problem. Normally, breaks in the oxide skin of aluminum heal instantly when the exposed metal reacts with, say, air or water. But if, for example, mercury or certain alkalis or acids dissolve this skin, the exposed underlying metal reacts vigorously. So, while aluminum cookware and foil are safe and useful in the kitchen, it is important to keep them away from strong salt solutions or caustic soda, for example, and also from wet food when it is not actually cooking.

This is because wet, fatty materials, such as cooked lard, form fat-soluble detergents that penetrate microscopic chinks in the oxide layer, exclude air that otherwise would reseal the skin, and corrode pinholes into the metal. If floating fat has coated the metal, even cold chicken soup can eat through a thick aluminum pot overnight.

The black stain is mainly from small amounts of iron in the aluminum. It is not deadly, but it is better not to eat food contaminated with high levels of metals, which also spoil the taste. For wrapping cooked fatty or acidic food for more than short periods, plastic film is much better.

Jon Richfield

THE SPRING IS SPRUNG

The mineral water in my local shop has a label telling me it is from a three-thousand-year-old source, yet there is still a "best before end" date on it approximately two years in the future. If the water has been in its aquifer for three thousand years, why should it go bad in a sealed bottle?

Lewis Smith

Mineral water has passed through layers of rock that have different effects on the water. Some minerals dissolve in the water, supposedly improving both its taste and health-giving properties, hence the demand for it.

The small pore size of the rocks that the water passes through acts as a filtration system, improving the purity of the water by removing larger molecules such as biological contaminants. As soon as the water emerges it is vulnerable to contamination again. The "best before" dates are based on the amount of time the bottler believes the water will remain without measurable levels of contamination due to the lack of completely sterile conditions in their bottling plants.

If the water is stored in a plastic bottle the date might also relate to contamination from the constituents of the plastic, which may change the taste of the water.

John Thompson

The reason for the "best before" date on bottled spring water is not the contents but the container. Most mineral or spring water is packed in polyethylene terephthalate (PET) bottles. During the manufacture of the bottles, traces of catalyst or plasticizer, which may include antimony, remain in the plastic and are leached out into the water over time. To avoid this, glass bottles, which have stood the test of time, are preferable.

Rob Davids

"Pure" water does not decompose or suddenly go bad. However, manufacturers of foods and beverages have to give "best before" dates to cover their backs. If the bottle sat around for long enough the plastic might decompose or the seal might degrade, allowing bacteria to enter and contaminate it.

As for the water being three thousand years old, in fact most of the water we drink has probably been in existence as water molecules for millions of years. What is important is the purity of the water, not its age: three thousand years in an underground aquifer may have filtered out all the organic matter, but it may still contain harmful dissolved chemicals such as arsenic.

Simon Iveson

TASTEFUL MATTERS

Why do cooked foods taste different after they have cooled from the way they tasted when they were hot?

Alan Parson

Cooked, solid foods are not static substances. Chemically and physically they are complex dynamic systems, continuously changing without stopping to suit anybody, so there are penalties for eating them too early or when they are past their best.

Cooking or cooling changes various substances in foods, affecting composition and flavor. Yesterday's leftovers have undergone reactions, including oxidation and the evaporation of aromas and flavor components. Food also changes physically on cooling, by congealing, crisping, or crystallizing, for example. These changes may prevent some substances from reaching the nose or the tongue, or expel or redistribute fluids. Few such changes reverse precisely on reheating, just as one cannot uncook food by chilling it.

Some changes are desirable, such as the setting of jelly or ice cream, but it is not for nothing that various foods are prepared at a particular temperature. Fresh hot food presents aromas in particular balances that reheating can never recapture.

Colin Collinson

Only a small amount of flavor is detected by the tongue, where taste buds recognize just five specifics: bitter, salt, sour, sweet, and umami (or savory). The majority of what we call "taste" is more specifically described as "flavor" and it comes from odor identified by nasal olfactory cells. That requires the flavor molecules to be wafted up from the mouth. This is more readily achieved when food is hot, creating convection currents and making odor molecules—as well as water molecules—volatile and mobile.

Water from food and saliva can dissolve flavor molecules so those more readily reach the taste buds while flavor vapors hit the nose. Having a cold that blocks the nose is even

more effective in reducing flavor than eating food cold, and can make apples and onions indistinguishable.

Elisabeth Gemmell

FLAT REFUSAL

Why are fizzy drinks such as cola or champagne far more appealing than the same liquid once it has gone flat?

Olaf Lipinski

Most fizzy drinks are made so by injecting carbon dioxide into the liquid at high pressure. Carbon dioxide dissolves readily at atmospheric pressure, but the high pressure allows even more to be dissolved. It forms carbonic acid in the drink, and it is this which gives the drinks their appealing "fizzy" taste—not the bubbles, as many people believe. When the drink goes flat, most of the dissolved carbon dioxide has been released back into the atmosphere, so the amount of carbonic acid is also reduced.

The fizzy taste is more appealing than the flat one simply because the drink was meant to be fizzy. Cola and champagne are concocted with the fizz in mind, using the carbonic acid as an essential ingredient in the flavor, so they will naturally taste better when the drink is still fizzy. When they go flat, this means that one of the main flavors has disappeared, and the overall taste will change—usually for the worse.

Martin Roos

A good taste is a matter of blended, often contrasting, sensations and expectations. These include temperature—for hot and cold drinks, say—sound and texture for chips or creams, plus aroma, flavor, and stimuli on the tongue. Fizz is generally created from carbon dioxide, though pressurized air also lends some noncommercial spring waters a certain liveliness. A good fizz tickles the nose and splashes minute stimulating droplets around the mouth as you drink.

Dissolved carbon dioxide has a distinct taste of its own, which is slightly sharp. Flat beverages have lost this bite. Going flat upsets the balance of the flavors and other stimuli, and without them such a drink is likely to taste insipid or too sweet, and . . . well . . . flat.

Jon Richfield

A side effect of taking the drug acetazolamide is that all carbonated drinks taste flat. Acetazolamide is used to help prevent altitude sickness by pre-adjusting the acidity of the blood to acclimatized levels. This also counteracts the acidity caused by carbonic acid in fizzy drinks, making them taste as if they were flat. I experienced this odd phenomenon firsthand last summer while drinking a soda before climbing Tanzania's Mount Kilimanjaro.

David Clough
University of Cambridge, United Kingdom

NO WRINKLES

How do they get the smooth, round chocolate coating on confectionery like Whoppers?
BBC Radio 5 Live listener

I spent six months making Smarties, a similar type of confectionery, in 1977. The chocolate centers were tumbled in a device resembling a cement mixer that gave them repeated coatings that alternated between sweet starchy liquid and powdered sugar, blow-dried after each coat. It took a week or two to learn the knack of ensuring an even coating: we had to remove clumped material, get the right combination of wet and dry, and keep the layers thin. Trainees' lumpy sweets were sold off cheap.

I handled about a ton of chocolate centers a day, putting on the white inner coat. More experienced workers did the outer candy coating, in similar "cement mixers," and the finished product was polished by tumbling the sweets in powdered beeswax, except for the black ones, for which petroleum jelly was used, apparently to avoid a whitish bloom.

Significantly, this was not a conveyor-belt manufacturing process. Each worker controlled their own rate, taking anything from an hour to an hour and a half per batch, depending on experience.

Peter Verney

MINIMUM DAILY REQUIREMENTS

I have heard that a family of four can be kept fed 365 days a year using only about 9.5 square yards of land. Is this really possible anywhere in the world? Could it really take only two hours a week as was suggested, and what would be on the menu?

Jan Horton

Opinions differ. There may be no definitive answer until somebody measures the output from 9.5 square yards of land—an experiment which is of necessity almost certainly going on in many poor countries.—Ed.

Energy flow is a key issue. The sun's intensity at the Earth's surface depends on latitude and season. The average value over a twenty-four-hour period across the whole of the Earth's surface is about three hundred watts per square yard. Therefore each day, a 1-yard-square plot receives an average of about twenty-six megajoules of energy—more close to the equator. The recommended dietary intake is about two thousand kilocalories a day. So, in theory, an average 1-yard-square plot receives enough solar energy to support three people. However, photosynthesis has an efficiency of only 10 percent so you would need more than 3.5 square yards per person. The figure of 2 square yards per person might just be achievable near the equator, although this seems optimistic.

There are also difficulties in getting the required nutrients and minerals, and in seasonal reductions in output.

Simon Iveson

I don't weigh my garden produce, but this year I did grow enough to fill a freezer, plus the produce my family ate fresh. All of this was grown on two small patches of land totaling about eight square yards. I believe I could have grown the minimum daily requirements for two, or possibly even four, if that had been my intention.

The produce—interspersed and rotated—included runner beans, sugar snap peas, onions, parsnips, raspberries, strawberries, spinach, broccoli, cauliflower, and blackberries. I grew carrots, tomatoes, cucumbers, peppers, zucchini, and herbs in pots on a one-square-yard shelf in my greenhouse.

I grow more intensively than advised by seed packets, and I start most of the outside crops in a heated greenhouse in winter. Some crops, such as beans, take up very little ground space, and crop rotation makes good use of space. In

addition we eat wild fare, such as rabbits, and we could have supplemented our diet in various other ways had we not preferred to encourage the wildlife rather than eat it. Two hours a week on a plot this size is plenty of time.

But could it work anywhere? The soil in my garden has been cultivated for generations. I recently started a vegetable patch in an uncultivated part of the garden and the result was poor. And I'm not sure I could have grown enough to feed us out of season without a greenhouse or freezer.

Tony Holkham

My family has decided that it would be possible to feed a family of four from 9.5 square yards of ground for a year if we were only producing vegetables.

You can grow climbing beans up poles along the rear of the plot and freeze the surplus. You can also grow trailing plants, such as pumpkins or cucumbers, within the plot, but let them trail outside it. Silver beet can be cropped continuously and potatoes can be grown in a stack of old car tires. Similarly, tomatoes and Brussels sprouts grow upward and you can bottle or freeze surplus tomatoes.

Herbs can be grown in pots with multiple openings, as can strawberries. Carrots, parsnips, rutabaga, and turnips can be grown between the tall plants. You need to stagger the plantings a little and freeze any surplus. Radishes are fast growing, so they need little space at any time, while celery is a "narrow" plant and the surplus can be frozen.

Keep seeds each year and store or barter the surplus seeds or grown vegetables for goods to preserve.

My garden is a little bigger than 9.5 square yards, but I haven't had to buy green vegetables (or eggs) for a family of three for longer than I can remember. In our case we also have chickens, which fertilize the soil and enter the equation themselves because they provide food (but take up space).

The vital part of this equation, however, is growing vegetables that can be stored or preserved.

Sandra Craigie

The difficulties inherent in calculating the food output of land are shown by the fact that our first correspondent above, Simon Iveson, later revised his calculations:

Since answering this question, I have thought of two important additional points.

First, the human body is not able to metabolize 100 percent of the energy stored in the plant material that it eats, so this would increase the land area needed to feed a person. Presumably the exact percentage we can metabolize depends on food type.

Second, cloud cover would reduce the amount of direct solar radiation that reaches the Earth's surface, further increasing the land area needed per person.

Two square yards is starting to look unfeasible.

Simon Iveson

AS TIME GOES BY

Why does red wine become lighter in color as it ages, but white wine become darker?

Volker Stuck

Color maturation in wines is just one small aspect of a very complicated chemical process. When red wines age they gradually turn from a deep purple color to a light brick red. Red wines are kept in contact with the grape skins throughout fermentation. During this process, blue/red-colored phenolic compounds called anthocyanins leach from the skins into the wine. As the wine ages, small amounts of oxygen react with anthocyanins and other, mostly colorless, phenolic compounds, causing them to polymerize and form pigmented tannins. Over time, these produce the brick-red color. Often tannin complexes grow as they react with other wine constituents, such as proteins, and many become too large to stay in solution and precipitate out, leading to the sediment you may find in aged wines.

White wines start out in bottles with a greenish tinge (young wines in Portugal are called *vinho verde*) and end up with a browner hue. White wines are not fermented with the grape skins, so they contain vastly lower levels of phenols, and therefore tannins. Also, white grapes do not contain anthocyanins—otherwise they would not be white. Consequently those few tannins found in whites are nonpigmented. It is presumed that white wines become browner with age because of the slow oxidation of what few phenols are present. A similar process can be observed in the discoloration of a half-eaten apple.

One interesting side note is that anthocyanins are only found in the skins, so it is possible to make white wine from red grapes if the skins are removed. White zinfandel, common in the United States, is an example of this.

Oliver Simpson

It was stated in a previous answer that "young wines in Portugal are called *vinho verde*." This is in fact incorrect. *Vinho verde* is a wine made with certain types of grapes in a certain region of Portugal, and there are white and red *vinhos verde.*

Although *vinho verde* should indeed be drunk while young (with the possible exception of *alvarinho* styles), the name itself does not imply youth.

Antonio Brito

QUESTION OF TASTE

Why does garlic make your breath smell in a way that, say, lettuce or potatoes do not?

Chris Goulding

Garlic produces a potent antifungal and antibacterial compound called allicin when the clove is cut or crushed. This is created by the enzyme alliinase acting on a compound called alliin. Allicin is responsible for the burning sensation you experience if you eat garlic raw.

However, allicin is not stable and generates numerous smelly sulfur-containing compounds, hence its pungent smell. After ingestion, allicin and its breakdown products enter the bloodstream through the digestive system and are free to leave again in exhaled air or through perspiration. This is the first effect of garlic.

In addition, the chemicals in garlic change the metabolism of the body and trigger degradation of fatty acids and cholesterol in the blood: this generates allyl methyl sulfide, dimethyl sulfide, and acetone. These are all volatile and can be exhaled from the lungs, giving you garlic breath the morning after a meal. It is not necessary to eat garlic to have garlic breath because allicin can be absorbed through the skin. Just

rubbing garlic on the surface of the body can be enough to generate smelly breath because it exits the body via the lungs.

The only real solution to smelly breath from garlic is for us all to eat it.

Peter Scott
School of Life Sciences
University of Sussex, United Kingdom

Garlic owes its pungency and subsequent halitosis-producing qualities to a variety of sulfur-containing compounds that are produced after cutting the cloves, some more transient than others and with a variety of health-giving properties. Sulfur is responsible for some of the smelliest substances known, from the brimstone stench beloved of vulcanologists and the rotten-egg smell of hydrogen sulfide to the potent secretions of the skunk.

Geoffrey Chaucer made a comment on alchemists of the fourteenth century in his own inimitable way:

> Evermore where that ever they gone
> Men may hem ken by smell of brimstone;
> For al the world they stinken as a gote . . .

Paul Board

CHEESE SCRUB

Mold has always been a menace on my blocks of Gouda and Edam cheese, which I store under a cheese cover. Recently my wife told me to put a lump of sugar under the cover with the cheese, and I have not seen mold since. The sugar gets moist

and slowly dissolves, but nothing else seems to happen to it. My wife learned this from her mother, and so it must be an old and possibly widespread remedy. Why and how does it work?
Georg Thommesen

This habit is also common in northern Germany, where the explanation given is quite simple. The sugar lump takes up moisture from the air trapped under the cheese cover, slowly dissolving as it does so. The relatively dry air reduces the suitability of the environment for cheese molds.

David Fleet

The sugar cube absorbs water, lowering the relative humidity, so that mold can no longer grow on the surface of the cheese. Salt would work just as well, as would saturated solutions of sugar or salt—saturated solutions are those that still contain some undissolved sugar or salt.

This forms the basis of humidity control in museum display cabinets. If the humidity is too high, undesirable molds grow, but if it is too low, wood and leather might crack. Saturated solutions of different salts can peg the relative humidity anywhere from 10 percent to 90 percent. For example, a saturated solution of lithium chloride will maintain a relative humidity of 11 percent, while a saturated solution of common salt keeps the relative humidity at around 70 percent.

John Hobson

The sugar will draw liquid from the air by its intrinsic hygroscopicity—its tendency to absorb moisture. This is the reason sugar cakes in the damp, and in the process it will suppress the growth of mold or bacteria.

The mechanism is related to the one that protects honey

from microbial growth. Honey is so effective at this that it was once spread on wounds to prevent infection. Honey suppresses mold and bacterial growth thanks to its high concentration of sugar. By drawing water away from its surroundings, the sugar desiccates any fungal and bacterial cells and spores in the honey. Cells must feed to reproduce and so they absorb food in contact with their cell membranes or which their excreted enzymes have released. Sugar in the honey will draw water out of the cell, which will either kill it or encourage it to live on in the spore phase and eschew reproduction until it encounters a more benevolent environment. This is the spore's job, and so it sits and waits and ceases to be active in the honey.

Bill Jackson

SHAPING THE MOLD

I discovered a pear that had started to go bad in my fruit basket. The first evening it had a perfect bull's-eye pattern of mold. Sixty hours later it had grown more (partial) rings of mold. Another forty-eight hours later it had grown still more partial rings, always separated by the same gap and all still roughly concentric. At that point it was getting pretty rotten, so I threw it away. But what causes the mold to grow in rings? I've seen it in other similar fruit.

Bob Ladd

The pear is suffering from brown rot disease, which is caused by the pathogenic fungus *Monilinia fructigena*. This is a very common and widespread disease of apples, pears, and stone fruits and spreads through the air as spores. The spores germinate on areas of damaged fruit, attacking it where the fungus has easy access to the unprotected, nutrient-rich, fleshy parts inside.

The fungal threads, or hyphae, grow and branch within the tissue and degrade the flesh. At first, the disease is invisible to the naked eye, but as it spreads, the pear responds with the typical "browning" reaction that gives the disease its name.

As it grows, daylight prompts the fungus to produce more spores on specialized hyphae that grow back out of the skin, forming gray-brown pustules.

A new crop of fungal spores is therefore produced with each period of daylight, and the fungus continues to grow as the flesh forms successively larger rings each day, giving the typical appearance described by the questioner.

A parallel situation can be seen in the "fairy rings" of dense, green grass growth and toadstools that appear in lawns—again it is a visible manifestation of a microscopic fungus growing beneath the surface. In this case, however, it is the fungus breaking down organic matter in the soil which causes the release of nutrients to stimulate grass growth and provide the essential energy to form the spore structures of the fairy-ring toadstool.

Peter Jeffries
Technology and Medical Studies
University of Kent, United Kingdom

PULLING POWER

It is accepted among Australian beer drinkers that a glass of cold draft lager holds its head for longer if the beer is pulled in

two or three separate pours rather than in one continuous pull. What causes this effect and does it also apply to beers served at cellar temperature?
Justin Swain

If you pour the beer in one pull, the foam grows under uniform conditions, producing relatively few bubbles and mostly large ones. Large bubbles pop quickly, so the head doesn't last. By pausing during pulling, one gives the first bubbles time to grow larger and more flexible before the turbulence of the next pull shears some of them into more numerous, smaller bubbles. Furthermore, the concentration of carbon dioxide in the poured beer surrounding a large bubble has had a chance to drop, so that the bubbles stop growing so rapidly. The effect of the extra pulls is to reduce the size of the bubbles and increase their number. This means a smoother, finer, firmer froth. Because smaller bubbles do not pop as easily as large ones, the finer froth lasts longer too.

Qualitatively, warmer beer behaves in much the same way, but it froths too violently and briefly, which masks the effect. Bubbles in warmer beer are larger and more fragile anyway, so the overall improvement is less worthwhile.

Billy Gill

KITCHEN CALAMITY

Once when I was reducing red wine and olive oil in a flat frying pan on the stove, the mixture exploded with an audible pop, spraying the wine—but not the oil—up to two yards away from the pan. Significantly, the wine was not hot enough

to scald me. I hadn't stirred the pan for at least a minute, and the wine and oil had separated. What happened?
Alan Gammons

There are several ideas about what could explain this. If any reader feels the urge to find out for themselves which scenario is correct, please exercise extreme caution.—Ed.

I don't know if chefs have their own term for this phenomenon, but chemists call it "bumping." It occurs when a liquid is heated above its boiling point but does not boil due to a lack of nucleation sites—scratches, sharp corners, or solid particles where bubbles form easily.

You can observe nucleation sites in a flute of champagne. They are the spots on the inside of the flute where bubbles are continuously forming before streaming upward. They usually mark the location of a tiny scratch or a speck of dust: a very clean flute, with few or no nucleation sites, will keep your champagne fizzy much longer.

When there are no good nucleation sites in the frying pan, the temperature can rise well above the liquid's normal boiling point, until it is so high that bubbles can form even without a nucleation site. Once a bubble forms, however, it acts as a nucleation site for the adjacent liquid. When such superheated liquid suddenly finds itself adjacent to a nucleation site, it boils explosively. This often happens in the chemistry lab because chemical glassware is so clean and defect-free that it often has no nucleation sites at all.

To prevent bumping, a chemist might deliberately scratch the inside of a flask to create nucleation sites, or may add chemically inert "boiling chips" to a solution. In other cases,

bumping is a necessary evil and the chemist tries to catch or deflect the resulting spray using a "bump trap"—usually in the form of a protective glass screen.

Nucleation is an important concept in many disciplines. In meteorology, for example, seeding clouds involves providing nucleation sites for the condensation of water vapor. And in metallurgy, the way that atoms crystallize into grains around nucleation sites greatly affects the strength and other properties of a metal or alloy.

In the case mentioned above, my guess is that the wine in the frying pan bumped. It may be that oil stuck to the surface of the pan, creating a completely smooth surface, free of nucleation sites, or it may be that the pan was simply very clean and smooth. The wine was presumably hot enough to scald when it left the pan, but tiny droplets flying through the air cool rapidly. To prevent bumping in your kitchen, you might provide some nucleation sites, perhaps by dropping a sprig of rosemary into your oil and wine before heating it.

Ben Haller

The mixture had not been disturbed and so had separated into oil on top and wine on the bottom, next to the heat. The alcohol in the wine would boil off well before the water in the wine could reach its boiling point and would bubble gradually through the layer of oil. Being heavier than air, the gaseous alcohol would sit on top of the oil, somewhat contained by the edges of the pan and mixing with the surrounding air to form an explosive mixture. After a while this mixture would overflow from the pan and slip down into contact with the heat source, igniting it, and the whole gaseous mixture above the oil. The resulting explosion would send a shock wave in all directions, causing the audible pop. The shock wave traveling downward would hit the viscous oil layer and force it downward too, pushing on the wine, which would have nowhere to go except sideways, up the sides of the pan and out over the kitchen.

David Levien

It looks like the red wine became superheated and flash-boiled explosively. Liquids start to boil when their vapor pressure equals the ambient pressure—in this case it was one atmosphere above the open pan. Usually the heat is released relatively smoothly, but two-phase liquids, such as this mix of oil and wine, boil at a lower temperature than either liquid, or phase, would boil on its own. Each phase generates its own vapor pressure and the mixture boils when the combined vapor pressure reaches ambient pressure. However, if the mix is allowed to separate before getting hot, as this one did, the temperature of the phase below—the red wine—can then exceed the two-phase boiling point before the red wine has even started to boil. When the wine does boil, it causes remixing, which results in a swift drop in boiling point and a rapid or instantaneous boil-off.

The wine was expelled rather than the oil because it provides almost all of the generated vapor. It was probably also hotter than the oil but fortunately hit your skin at a cool temperature because it was probably dispersed in a rapidly cooling aerosol by the force of its vapor expansion.

Paul Gladwell

2 DOMESTIC SCIENCE

WHAT'S IN THE BOX?

Whenever I empty a carton of a drink such as milk or orange juice, I replace the cap on the container and leave it aside ready to throw out. Invariably, when I return the carton seems to be under some pressure as the sides are bulging and stiff. Why does this occur? This does not seem to happen with cartons that have held carbonated drinks.
Cheryl Altschuler

Two possibilities exist for this effect. We outline both below plus an additional phenomenon suggested by another reader.—Ed.

When you finish the last of the milk or orange juice and put the top back on, most of the air in the carton is still chilled from being in the fridge. As the air reaches room temperature it expands, swelling the container. Try leaving the empty container with no top on until it reaches room temperature and then put the lid on—you'll find the container won't swell.

You can produce the reverse effect by washing out the empty milk container in hot water, then immediately putting the lid on. The container will now suck in its sides as the air inside it cools.

Carbonated drinks containers, on the other hand, are designed to hold their contents under pressure. If you have been sprayed by a fizzy drink you will be able to vouch for the pressure such containers can withstand. When the cool air expands to room temperature in an empty carbonated drinks bottle the pressure on the more rigid structure is well within its design constraints and it does not visibly swell.

Mark Edwards

Fruit juice and milk cartons bulge through accumulation of fermentation gases from a range of microorganisms. These are thriving on the nutritional soup left behind in the cartons. Long-life drinks are usually pasteurized during processing by heating them sufficiently to kill most spoilage organisms. As soon as you open the container, however, bacteria, yeasts, and microbial spores will contaminate your product.

If you keep the carton in the fridge, the growth of these organisms will be slow. However, once you leave an empty carton at room temperature the organisms multiply quickly on the film of juice or milk left inside. They will generate enough carbon dioxide to create a discernible positive pressure, especially in warm weather.

Fermentation is much less likely in carbonated drinks because microbial inhibitors such as sodium benzoate or potassium sorbate are often added. Furthermore, the concentration of nutrients is often low, especially in sugar-free drinks, because these products contain little or no real fruit juice.

The organisms you might find in your fruit juice cartons are typically types of yeast, such as species of *Saccharomyces*

or *Candida*, which thrive in the high-acidity environment of fruit juices.

Milk is a great food for most microorganisms, and many types of bacteria can proliferate in it. Even if milk is not contaminated from the outside, fermentation may still occur thanks to the growth of bacilli, whose spores are resistant to the temperatures used in milk pasteurization.

Julia Aked

You can have similar fun at breakfast time by taking a cold bottle of milk—the kind with a foil cap delivered to the door—from the fridge, pouring a little onto your cereal, and replacing the cap securely. A little while later, when the air in the bottle expands as it warms up, the foil top will shoot into the air with a quite impressive pop.

Presumably this phenomenon also explains why, having taken my baby's milk from the fridge and warmed it, when I come to test the temperature by inverting the bottle over my wrist, the milk shoots all over the kitchen unless I first unscrew and rescrew the lid to release the pressure.

Claire Webster

LASTING STRIPES

How do toothpaste makers get the stripes in toothpaste? And why do they persist until the tube is used up?

Miles Ellingham

This is one of those simple inventions that has been helping manufacturers to sell toothpaste for decades. We have to go back almost fifty years to find U.S. patent number

2,789,731 and UK patent 813,514, both in the name of Leonard Lawrence Marraffino. He licensed his invention of striped toothpaste to Unilever, and this company subsequently marketed the first commercial version, which was called Signal in the United Kindom and Stripe in the United States. When the tube was squeezed it produced red stripes on white toothpaste.

Behind the nozzle was a hollow pipe that extended a little way back into the toothpaste tube. The white paste traveled down this pipe. Around this pipe was the funnel-shaped neck of the tube and, at the nozzle end of this, the pipe had tiny holes that opened into its interior. When filling the tube, red paste was first squirted into the funnel-shaped neck until just shy of the rear entrance of the pipe. Then the white toothpaste was added, and the end of the tube sealed.

When you squeezed the tube, the white paste flowed out through the pipe, but it also compressed the red paste, forcing it through the tiny holes into the pipe to form the stripes. If the stripes stopped coming out, they could sometimes be made to start again by warming the toothpaste tube in hot water.

Colgate-Palmolive's U.S. patent number 4,969,767, granted in 1990, describes a scheme for adding stripes of two colors. Here, the central pipe is surrounded by a wider but shorter pipe, which creates a space for a second colored paste. Tiny holes once again deliver this paste onto the surface of the white paste.

Tom Jackson

Another way to create striped toothpaste is to pack columns of different colored pastes side by side in a vertical dispenser. The pattern is reasonably well preserved even when its diameter is reduced as the pastes are squeezed and forced through the hole at the top of the tube. However, unless you use a pump dispenser, the action of squeezing the tube tends to mix up the stripes.

Neil Rashbrook

Thanks to Blair Scott of Falkirk, United Kingdom, for spotting an interactive demonstration online of pump dispensers being filled using this latter method (see http://aquafresh.co.uk/KIDS/activities/milk-teeth-activities/how-do-we-get-the-stripes-into-aquafresh.aspx). Patents of the inventions mentioned by our first correspondent, complete with diagrams, can be found at http://www.google.com/patents.—Ed.

MONEY BAGS

When my bank debit card doesn't work when swiped, some gas stations seem to have an ingenious and effective solution. They wrap the card in a thin, clear plastic bag and swipe it again—and it works. Why?

Matt Huddlestone

The cardreader has a small induction coil that detects a succession of magnetic and nonmagnetic zones in the card's magnetic stripe. When the card is pulled through the reader, each magnetized zone makes a small electrical pulse as it passes the coil. The zones are arranged to encode the data needed to complete a transaction—along with, in countries such as the United States and the United Kingdom, the client's PIN (personal identification number).

The magnetic stripe consists of magnetized particles embedded in a plastic binder with lubricant. It can be damaged in many ways, including exposure to a large external magnetic field, whether from a permanent magnet or a high-strength alternating current. Damage can also occur by a gradual "smearing" process, in which some magnetic particles are physically dragged from one part of the stripe to another. This transfers a few magnetized particles into zones of the stripe that previously had none, corrupting the data. This is by far the most common mode of failure, and this smearing process is probably under way on your card.

However, the "contaminated" zones still have a lower magnetic field than the properly magnetized zones. Magnetic field intensity is governed by an inverse-square law, so if the card normally passes within 0.01 millimeters of the reader and you double this, the signal picked up by the reader will fall to a quarter of its normal level. Just a small increase in distance, such as that provided by the plastic bag the questioner has noticed, will reduce the weak signal from a contaminated zone—which is supposed to be blank—to a level that the reader is likely to register as a zero. This means the card will work properly again. This extra distance can be created using any nonmagnetic spacer, such as the plastic bag or sticky tape.

Cardreaders also vary in their sensitivity. Some will read the card properly, and others will need the bag or tape. This situation will progress after an initial failure until the remedy stops working. So, once you have experienced the first occurrence, arrange for a card to be reissued.

Bill Jackson

WINTER CHILL

If we set our refrigerator thermostat to keep the contents at a reasonable temperature in summer, liquids turn to ice overnight in winter. This has happened with every fridge we have owned. Why doesn't the thermostat react only to the temperature inside the refrigerator? Does something similar occur in electric ovens?

Agnes and Puka Henry

Domestic appliances controlled by thermostats do not flip on and off at precisely the set temperature. Instead, they have a range of tolerance. Modern fridges in particular tend to be designed with miserly insulation in mind, so if their thermostats were too precise they would have to switch on and off every minute or so, decreasing the life of the thermostats and of the refrigeration units themselves. The average temperature in a typical fridge is about 39°F, but during the cooling phase it may drop below 32°F, afterward warming up to 42°F or thereabouts.

In high ambient temperatures, the fridge soon warms up enough to switch the thermostat on again, but when the weather is colder the low temperature phase may persist long enough to freeze the contents. If this is a nuisance, set the thermostat a little higher; just don't forget to reset it when the weather warms up.

In most regions, fridge temperatures are only a maximum of 68°F below ambient, so such seasonal changes are quite marked. Although similar principles apply to oven thermostats, oven temperatures typically are so far above ambient that seasonal temperature variations hardly affect their operation.

Jon Richfield

Your fridge may suffer from problems that have their root in minimalist design philosophy. In a perfect world your fridge would have very good insulation and a large enough heat pump for all conditions. It would also have one cooling coil for the freezer portion and another for the refrigerator area that stays above 32°F. All this costs money, so on many cheaper fridges the insulation is thinner than it should be and the heat pump's compressor is smaller.

Some fridges also use one cooling coil for both the freezer and refrigerator sections and simply duct some air from the freezer to the refrigerator section. The thermostat is located in the freezer and the user turns a vent dial to control the amount of cold freezer air allowed into the refrigerator section.

As you might expect in such models, the difference in heat flow through the walls in the winter and summer means that the contents of the refrigerator portion can freeze in winter and may become too warm in summer. The solution is to change the setting as the ambient temperature changes. And when you replace your fridge, give a thought to the economics of a more efficient fridge.

More costly fridges do have thicker insulation and separate cooling coils for each section. Some of the most efficient fridges have a separate compressor for each section. If you have a fridge that is more than fifteen years old, you would save money by throwing it away and buying one with dual compressors and ten-centimeter (about four-inch) insulation, rather than the more standard five centimeters (about two inches). These fridges last twenty years and save you about $13 a month for that period, a total of $3,120. I laugh when I see people with a really old fridge in the cellar which they use to chill beer, consuming a thousand watts at

a cost of about $35 a month. You could buy a lot more beer with that.

Bill Jackson

CUT-GLASS ACCIDENT

I recently removed some crystal glasses from a cupboard. Two of them left about half an inch of their rims behind on the shelf. The glasses are more than thirty years old and were upside down. The rings of glass are not an even width, and there appears to be a starting point for the "cut." What caused this?
Linda Sulakatku

Perhaps no question has caused as much consternation in New Scientist *as this one. At first we were convinced by what later turned out to be an ancient urban myth when we concluded that glass can flow like a gel. Eventually, after running a number of theories as possible answers, we were put straight by John Parker of the University of Sheffield. The answer which proved erroneous is included here as a monument to our shame and as evidence that it is the job of science to question itself again and again.—Ed.*

Glass looks and behaves like a solid, but it is actually a gel—a very slow-flowing one. The structure of lead glass or crystal causes it to flow more readily than other types of glass. You might have seen this in windows in old buildings, where the glass is visibly thicker at the bottom than at the top. Lead glass, from which these crystal glasses are made, also becomes more brittle over time.

The base and the stem of the glasses are quite thick, to

provide a base that is heavy enough to stop them falling over when you use them. The sides of the glasses are elegantly thin.

While the glasses were stored rim down, the glass will have been slowly flowing toward the rims, making them fractionally thicker. The rest of the glass will have become slightly thinner, and all the glass more brittle. While the glasses were upside down, the force on the rim was compressive—with the weight of the heavy base compressing all the glass below it. When the questioner lifted them, they changed this to a tensile force. After long storage the sides had become too thin and too brittle to support the weight of the rim, and the glass broke.

Slight variations in the thickness of the glass to start with mean the break will not be even. Lead glass has impurities, and one of these probably provided the point of weakness that started the break, which propagated from there.

Cold comfort I'm sure, but if the glasses hadn't broken when you picked them up they would have broken when you filled them, and showered you into the bargain.

Helen Jenkins

The body of the glass is blown as an egg-shaped globe or bubble and the stem and foot are added while it is still hot and malleable. After cooling, the rim line is scratched round and the cup part snapped off, leaving a sharp cut edge.

This edge is then reheated, and it flows in a treacle-like state to form a smooth, rather than a sharp, rim. Finally the whole glass is heated in an annealing oven to soften it slightly and allow the stresses that have built up within the glass due to heating and cooling to relax. Then the whole object is cooled evenly.

In this case the annealing process was probably not carried out correctly. If the glass has a scratch, it can create a stress line around which a crack can easily run. Such failures can occur spontaneously: in the late 1940s there was a batch of toughened glass tumblers on the market that were prone to exploding spontaneously.

David Stevenson

This reminds me of a curious incident at a pub in which I used to work. Occasionally, when we were pouring beer into pint glasses from the hand pumps, the glass would break into two parts cleanly around the middle. It left a perfect cut and a clean edge—and, of course, beer all over the floor.

This happened several times, so we examined the glasses on the shelves. Many of them seemed to be scored as if by a glass cutter. We assumed for a while that a customer was playing some sort of prank. Eventually the landlord had a flash of inspiration and asked whether any of the barmaids was left-handed.

One confirmed that she was, and she wore an engagement ring which carried a large diamond. When she dried pint glasses she held the cloth in her left hand, inserting it into the glass and rotating it to dry the inside. The diamond acted exactly like a glass-cutting tool and left a score mark all the way round that caused the glass to neatly divide under the shock of beer hitting it.

I wonder if we may be seeing a similar situation with your reader's crystal glasses.

Richard Hooker

As we said above, readers of New Scientist *really beat us up after we repeated the myth that glass flows. So we sent all the original answers to John Parker, in the glass science and engineering department at the University of Sheffield.—Ed.*

The contention that glass flows is certainly wrong. From what we are told, the glasses had stuck to the shelf on which they were placed, so when the owner tried to remove them this induced sufficient stresses to cause failure. Two questions need to be considered: why did the glasses stick, and why did that particular failure occur?

An examination of the marks left on the shelf after removing the glasses might have provided some clues. While glass does not flow, it is very slowly attacked by liquid water and the reaction products might just have produced a weak bond with the shelf over time. Such bonding can be a serious problem in hot, humid climates if glass sheets are stacked together and condensation and evaporation occur alternately. Even after a short period in storage the sheets can become bonded. It would be interesting to know whether the glasses were put away in the cupboard wet, or whether conditions were so humid that further attack occurred over a long period.

If the shelf itself had a plastic coating this may also have contributed. Plastics do change over time as a result of flow, reaction with sunlight, and so on. It is even possible that an airtight seal formed between the glasses and the shelf so that a difference in air pressure contributed to the adhesion.

The cleanness of the break suggests that the crack traveled around the glass as a result of quite low applied forces. Under these conditions the fracture surfaces tend to be smooth and growing cracks do not fork. The crack will have started at one point, near the back of the glass, where the action of pulling on the stem induced a bending stress that

was tensile on the outside glass surface. Examination with a hand lens might have allowed the origin of failure to be identified. Once a crack has started it will run around the whole rim very easily.

But why did the crack start? Either because the applied stresses were locally high or because a flaw in the glass surface at a critical point boosted low applied stresses. This could have been caused by a thin region in the glass or because poor annealing had introduced extra stress. However, a severe flaw—for example caused by a scratch from a diamond ring as mentioned by one of your correspondents—can boost stress levels at the tip of the crack by orders of magnitude in a brittle material such as glass (a brittle material being one that does not flow under stress). A defect like this is the most likely explanation.

As always with these kinds of problems the danger is that only part of the story comes out in the first analysis and another factor turns out to be the key.

John Parker
Department of Glass Science and Engineering
University of Sheffield, United Kingdom

WHAT A BRIGHT SPARK

My son was playing with sparklers one night and wanted to know why he could draw shapes in the air with them. Then he wanted to know why each "sparkle" that shot off the wire he was holding consisted of a distinctive line ending in a star. I couldn't answer either question. Can anyone help?
Simon Zia

Sparklers appear to draw lines in the air because of the phenomenon known as visual persistence. The human eye does not react instantly when its view changes, but keeps the old image around for a few milliseconds. This is what enables us to perceive films or television images as moving pictures when they are in fact a sequence of still images. The persistence of the eye causes each image to merge into its successor, creating the illusion of movement.

If the changing image contains very bright objects against a dark background—such as a sparkler at night—the persistence lasts longer, so the light from quite a long period of time can be added together to appear as a single streak.

There are numerous gadgets that exploit this effect by using strips of fast-moving LEDs to apparently create writing in the air. Persistence can also be seen in the colored spots left in your vision after a camera's flash has gone off.

The sparks from the sparkler are produced by burning flecks of a metal such as magnesium or aluminum flung off from the firework. Initially only their outer layer burns, but after the fleck has burned down to a critical size the core becomes so hot that it explodes. The subflecks from the explosion then burn out quickly and brightly in a distinctive star.

Alec Cawley

WHAT'S THE ATTRACTION?

If I move an iron nail slowly nearer to a magnet, at some point the nail will leap forward and stick to the magnet. I understand enough science to know that there is no such thing as a

free lunch, so where does the energy come from to overcome the inertia and friction, and move the nail?
Peter Jenks

Our one free lunch is that we live in a universe that offers usable energy. Imagine a rock falling to Earth from space. Its position above the planet plus the gravitational field gives the system potential energy (energy that is stored) that converts to kinetic energy (energy of movement) as the rock accelerates earthward. Had we lifted it into its starting trajectory from Earth, we would have had to spend, and the rock would have released, exactly the same amount of energy.

Similarly, the nail plus the distant magnet have potential energy that changes into kinetic energy as they "fall" toward each other under the magnetic field. The stone high above Earth and the nail far from the magnet are full of pent-up energy. It is the stone on Earth or nail at the magnet that has the lowest energy, not the stone in space or the nail by itself.

It makes no difference whether they started apart, or whether you spent energy pulling them apart before letting them come together again. The energy released equals the energy available, and the mass-energy in the universe is unchanged.

Jon Richfield

It helps to compare the magnet's magnetic field with the Earth's gravitational field. One might as well ask: "When you drop a ball, where does the energy come from to overcome inertia and air resistance, and pull ball and Earth together?"

The simple answer is: The energy was always there, stored as potential energy—the potential for the nail to fly toward the magnet always existed. When the nail was released, this potential energy was converted into kinetic, heat, and sound energy as it moved toward and impacted with the magnet. When you pull the nail away from the magnet, you exert mostly kinetic energy, which is converted back into potential energy for the nail, giving it once again the potential to be attracted to the magnet.

Sam Davies

Although the movement is sudden, the energy is there before the nail moves, stored in the magnet's magnetic field. What appears to be a sudden increase in energy is the change from magnetic potential energy to kinetic energy. The nail has potential energy at an infinite distance, but loses it as it is moved closer and the potential energy is converted into kinetic energy.

Potential energy is difficult to see, but if you imagine holding the nail and the magnet apart from each other in midair, you would have to pull the nail slightly to keep it away from the magnet. The closer the nail is to the magnet, the greater the force required to keep them apart. When you put the nail on a surface, friction holds it away from the magnet, rather than your fingers. At the point where the friction is less than the force required to keep the nail where it is, it will accelerate toward the magnet.

Adam Hewitt

SPIRIT IN THE SKY

How do manufacturers calibrate bubble levels? I have two bubble levels and one is way out. Can I recalibrate it?

Dave Gellard

The way I was taught to use a bubble, or spirit, level was always to check the level you are measuring twice, the second time by turning the level 180 degrees. This allows you to factor out any errors in the position of the bubble. If there are errors you will get different results, and to find the true level, you average out the error in the two measurements.

If you are trying to calibrate a level and have one that comes with a rotating dial for just this purpose, use a newly painted, smooth wall and fix a point of reference at one end. Now, place the level against the wall horizontally—according to the bubble—and draw a line from the point of reference to the level's other end. Spin the level 180 degrees, do the same again, and you will end up with two diverging lines. Split the difference at the divergent end, and you can scribe a line from this new point back to the point of reference that is perfectly level. You can then use this line to recalibrate your level.

By the way, I am also told that you can buy spare bubbles for levels. You'll find them on the same shelf as left-handed screwdrivers and long weights.

Marco van Beek

Where I live in Thailand the builders don't have levels. Instead they use a long piece of plastic tubing filled with water. When the tubing is allowed to hang in a shallow U shape and held against a wall, the surface of the water at each end will settle at absolutely the same level. The builders then use a "powder string"—a length of string covered in colored chalk—which is stretched between the two water levels, and with a swift twang a level line is drawn along the wall. You could hold your level up to this line to calibrate it.

Ken S.

Your correspondent should find a perfectly horizontal surface using a level that is known to be correct. Put the suspect level on the surface and adjust the angle of the bubble capsule. Finally verify the adjustment by turning the readjusted, suspect level 180 degrees as outlined in the first answer.

Andrew Fogg

3 OUR BODIES

THE BEST BITS

In a worst-case scenario, if one had to eat parts of oneself, which non-organs would be the most nutritious? Nails? Hair? Earwax?

David Klein

Nails and earwax are almost indigestible, unless you pressure-cook them for a long time, preferably in mild acid. If we take nutrition to mean energy or proteins and the like, then bone marrow or fat tissue would be the most rewarding.

Extracting marrow is a tough task, so using liposuction to harvest fat would be best, assuming you had some left to harvest because presumably if you were starving there might not be much. Fat tissue comprises about one-sixth to one-quarter of the mass of a vigorous person, and over three-quarters of a grossly obese one. But if you harvested too much at a time, the trauma could kill you. Drinking blood would not provide you with very much nutrition.

In truth, there is little nutrition on the typical human body available for nonsurgical harvesting. Unlike humans, scavengers like museum beetles and clothes moths can digest keratin—the protein of hair and fingernails—but about the only significant external item a human could consume would be their epidermis, soaked in water until some of it could be scraped off. This skin would probably be more nutritious if left to decay for a few days, so that bacteria could render it more cheesily digestible. Bon appétit!

Jon Richfield

Your body will start to consume itself in a reasonably efficient manner during starvation as fat reserves are used up. When the fat has dwindled the body turns to its muscle tissue.

To surgically remove tissue for consumption would inflict considerable trauma on a body already weakened by starvation. By cutting off bits you would reduce your chances of survival, since the damage to your body would instigate self-repair, which uses up yet more fuel.

Consuming your waste products such as urine would cause further problems. If you must drink urine in a survival situation, distill it first. Consuming contaminated brine when you already have so many problems is not a route to longevity.

Bimmo

I hesitate to suggest this, but you have already put forward earwax as an option (of which there would not be enough, even if it were nutritious), and your resistance to grossness may be weakening if you are already drinking your own urine in a bid to avoid dying of thirst. In light of this, might I suggest eating feces?

Feces contain lots of good protein because a major component is the dead bodies of a vast number of gut bacteria.

It smells bad, but holding your nose makes the taste bearable, apparently. I don't think eating your own feces could make you seriously ill, because you can only catch a disease such as hepatitis from the feces of someone who already has the disease. Similarly, although there are dangerous strains of *Escherichia coli*, if your own gut *E. coli* were dangerous you'd already be sick. To be on the "safe" side, you could cook the feces until their internal temperature reaches 160°F to kill off any *E. coli*.

For obvious reasons, however, I would advise getting professional advice before eating feces, unless your life really does depend on it. And don't forget to start collecting it before you run out of other food or you might not be producing much.

By e-mail, no name supplied

EYE LINER

Make a pinhole in a piece of cardboard. Bring your eye close to it and look through the pinhole as you move the card. You will see the network of your retinal capillaries against the background of a cloudy sky. How does this happen?

Doohan Cho

This is a fascinating phenomenon known as Purkinje shadows, after the Czech physiologist and neuroanatomist Jan Evangelista Purkinje. It also illustrates an excellent argument against intelligent design.

In the human eye, light passes through all the nerve fibers and blood vessels before reaching the photoreceptors.

This curious arrangement means that the blood vessels cast shadows on the back of your eye, and it explains why the capillaries can be seen if you look through a moving pinhole. Surely a master creator wouldn't have made a mistake like that. After all, the squid eye is designed the other way around, which raises the possibility that the mythical intelligent designer of life considers cephalopods a higher form of life than humans.

The reason we don't normally see the shadows of the blood vessels is because the human eye is incapable of registering a stationary image. We can see things that don't move, such as statues or doors, only because our eyes are continually making tiny movements which ensure that their images jiggle across our retina. Using sophisticated eye-tracking equipment, it is possible to completely stabilize any retinal image. When this happens the image disappears, a phenomenon called Troxler's fading. If our eyes were completely still we'd be almost blind. Intelligent design, huh? Because the blood vessels are part of the eye, they move with it—meaning they are essentially stationary as far as our photoreceptors are concerned—and therefore usually remain invisible.

Moving a pinhole across the pupil changes the direction of light reaching the back of the eye, which has the effect of "moving" the capillaries relative to the retina, making them visible. An even better way to see the blood vessels in your own eye is to put a small bright flashlight near the white part of the eye (while being careful not to poke yourself).

Ian Flitcroft
Consultant ophthalmologist

NO SWIMMING

Everyone I know was told as a small child not to swim within an hour of eating. Why is this?
Louis Counter

Several people suggested that this is a groundless old wives' tale, although our understanding of physiology suggests otherwise. A few pointed to carbohydrate loading, performed by marathon runners before a race, as evidence that the body copes well with eating and exertion after only a short interval. The question is, how short should that interval be? See Jon Richfield's answer below.—Ed.

I swam competitively throughout my school years and still maintain a fitness routine. I frequently consume a large meal and hop back in the pool with no noticeable effects during or after a prolonged casual swim. However, if I were to do a long-distance workout, I would find myself fighting uncomfortable reflux from my recently consumed meal. I've found that pickles are the worst. The problem seems to stem from the abdominal twisting that precedes shoulder rotation during freestyle crawl. Butterfly stroke doesn't seem any better—it feels like doing rapid sit-ups after stuffing yourself. Swimming on a full stomach is undoubtedly less efficient and markedly slower, although I've never experienced the cramping that some people report.

By e-mail, no name supplied

The admonition is certainly no longer taken very seriously, as many a postprandial, pandemonic children's pool demonstrates every summer. Still, our physiology does adjust radically to meet the challenges of exertion and digestion. Each requires resources to be concentrated on particular functions: we need a lot of blood to process food and transport it from the gut, and that competes with the need for blood to transport oxygen and fuel to muscles.

In nature, therefore, a full stomach inclines one toward torpor. Any need for heavy exertion at such times implies an emergency that may justify vomiting to jettison the burden. Snake handlers soon learn that their animals may regurgitate recent meals under stress, and human recruits under military training commonly leave their breakfasts by the roadside during heavy exertion.

One possible rational explanation for the idea is that because both exertion and digestion demand extra blood supply, swimming after heavy eating may be dangerous for anyone prone to fainting. Anyway, why risk vomiting during a strenuous swim instead of waiting a while? The stomach absorbs very little food, so it is an uncomfortable handicap when full, whereas food in the small gut continuously maintains your stamina while you assimilate it.

Jon Richfield

LIFE GLUE

What causes cells to stick together in the human body rather than simply fall apart?

McKenzie Gibson

Cells in the body are organized in tissues that are held together through a variety of molecular interactions. On the one hand, cells interact with each other. This is a very specific interaction mediated by various families of adhesion molecules called cadherins, neural cell adhesion molecules, and intercellular cell adhesion molecules. These are all expressed on the surfaces of cells and anchored in the cytoskeleton of each cell, an arrangement which stabilizes and gives strength to interactions between cells.

On the other hand, the body's tissues are not made up solely of cells, but of an intricate network of macromolecules too, called the extracellular matrix. It supports the assembly of cells and is composed of a great variety of polysaccharides and proteins, mainly produced locally by other cells called fibroblasts.

These macromolecules are combined into an organized mesh and, depending on the proportions of its components, the matrix can adopt diverse forms adapted to particular functional requirements. For example, it can be calcified and hard as in the bones and teeth, transparent as in the cornea, or elastic and strong as in the tendons. The main components of the matrix, which determine the properties listed above, are fiber-forming proteins that can be structural (collagens and elastin) or adhesive (fibronectin and laminin).

Cells adhere to this complex scaffolding through surface receptors called integrins that are anchored in the cell cytoskeleton and bind to the matrix components. Though integrins are densely packed on the cell surface, they have a relatively low affinity for interacting with the matrix components. This allows the cells to move within the matrix without losing their grip completely, meaning that it is, in effect, a rather flexible glue.

However, the interactions between integrins and matrix components have a deeper purpose than just holding the cells in place. Almost like antennae, they can transmit messages to the cell about the microenvironment to which it needs to adapt, and so influence cell shape, movement, and function.

There are also, of course, cells in the body that remain free. These are the components of the blood: red and white blood cells and platelets that normally float in the bloodstream, delivering oxygen to the tissues and keeping a lookout for invading microorganisms and wounds. These cells are capable of attaching themselves to other cells or tissues at will. For example, the encounter of a platelet with a wound activates the integrins of the platelet, enabling it to bind to fibrinogen in the blood vessel and initiate the aggregation process that forms clots and stems bleeding.

Alena Pance
Department of Biochemistry
University of Cambridge, United Kingdom

NASAL ILLUSION

If I tap my nose with my finger I only register a single touch, yet the sensation from nerves in my nose has only a few inches to travel to my brain, while the one from my fingertip has to travel about a yard up my arm and shoulder. Is this an illusion arranged by my brain, or is the brain unable to distinguish between two events so close together in time? Can anybody explain?
Geoff Lane

You do register two touches: the touch of your finger on your nose, and the touch of your nose on your finger. The illusion arises because you have many more sensory receptors on your fingertips than on your nose, and they are more sensitive in different ways. As an aside, look up "somatosensory homunculus" on the Internet.

If you stub your toe, the first thing you do is grab hold of the injured part with your fingers to see if it's OK. You use your fingers because they provide better information about that part of your body than the body part itself.

So you do feel two sensations. Your brain just chooses to disregard the sensory stimulation of the skin on your nose because it is a much less rich source of information.

Roy Hunter

The brain copes admirably with time lags, thanks to its internal workings that deal with the different reception times of incoming signals. It is so good at this that in some circumstances it is difficult for people to appreciate that a time lag exists at all. The brain seems to do this by tagging stimuli and thoughts with a time-stamp, though nobody knows how or what form this takes in the brain.

Additionally, interacting with other people requires making predictions of elements or events that you cannot yet have heard, seen, or sensed. The ability to sing in unison is a good example. You are committed to singing your note before hearing the same note from the other singers. Yet you believe the thought and the action coincide.

Experiments on response times within the brain show that the thinking and planning required for actions like singing a particular note start long before anyone can report they have decided to sing that note. The brain tags the thoughts into a single instance, so you—the internal observer of your own behavior—think that your thought, planning, and execution of the action all happened at the same instant.

To see that this is true, try very hard to work out when you planned the mouth and throat actions you need to utter the lyrics of a song you are singing. You'll conclude it all happened at the same time.

Christopher Cradock

NUMBSTRUCK

Why do we sometimes get "pins and needles," especially in our arms and legs, and what exactly occurs in our bodies to make this happen?

Kuna Patel

Pins and needles, properly called paresthesia, can be caused by a number of things. The most common cause is direct compression of a sensory nerve, inhibiting its ability to send sensory information to the brain. This often occurs where a nerve runs close to the body surface and on top of a bone. For example, compression of the ulnar nerve as it crosses the elbow causes pins and needles in the hand. This is why people wake up with pins and needles: they have compressed a peripheral nerve by lying on it.

Less commonly, abnormal plasma calcium levels can cause pins and needles, also by affecting sensory nerve function. Low calcium, sometimes seen after thyroid surgery, classically causes pins and needles in the hands. Hyperventilation also causes the sensation by making the blood more alkaline, thus driving calcium into cells and lowering plasma calcium. Pins and needles in the feet are a sign of nerve damage caused by poorly controlled diabetes.

Sinister causes of pins and needles include neurological

conditions such as stroke and multiple sclerosis. The short answer, therefore, is that pins and needles are caused by some factor impairing sensory nerve function.

David Anderson
Hawkes Bay Regional Hospital, New Zealand

Paresthesia describes a number of abnormal sensations, including pins and needles. Transient paresthesia is the temporary sensation of tingling, pricking, or numbness of the skin and has no apparent long-term effect.

It is commonly felt in the extremities and is usually caused by a lack of blood supply or by inadvertent pressure placed on a superficial nerve. For example, if you kneel or sit on your legs, the weight of your body tends to limit the blood supply to the lower limbs and, as a result, the nerves become starved of blood and start to send unusual signals to the brain. This is perceived as a tingling sensation or pins and needles in the foot or lower leg. Once you move and change position, the nerve compression is released and the pins and needles gradually fade.

Paresthesia may also be chronic. Poor circulation is common in older people. It may be caused by conditions such as atherosclerosis or peripheral vascular disease. Without a sufficient blood supply and hence nutrients, nerve cells cannot function normally. This is also why paresthesia can be a symptom of malnutrition, as well as metabolic disorders such as diabetes and hypothyroidism.

Additionally, inflammation of tissue can irritate nerves running through it, causing paresthesia. This is the case in conditions such as carpal tunnel syndrome and rheumatoid arthritis. Chronic paresthesia can sometimes be symptomatic of neurological disorders such as motor neuron disease (ALS, or Lou Gehrig's disease) or multiple sclerosis.

Melanie Trickett

LIFE AFTER DEATH

Why do hair and fingernails grow after death? Surely dead means dead. How can our bodies continue to produce more cells?

Shannon Smith

This is something that we noticed as fresh-faced, first-year medical students when confronted with the cadavers we were going to dissect over the next two years. All had slightly long fingernails, and all of the men had neatly cropped stubble. We assumed that these had grown while the cadaver was being prepared. However, an anatomy demonstrator assured us that nails and hair do not grow after death and that this phenomenon was actually the result of the surrounding tissue drying out and shrinking away from the nail folds and hair shafts, giving the impression of growth.

David Pothier

This is a myth possibly spawned by Erich Maria Remarque's novel *All Quiet on the Western Front* in which Paul Bäumer, the nineteen-year-old narrator, considers the death of his friend Kemmerich from gangrene. He writes: "It strikes me that these nails will continue to grow like lean fantastic cellar-plants long after Kemmerich breathes no more. I see the picture before me. They twist themselves into corkscrews and

grow and grow, and with them the hair on the decaying skull, just like grass in a good soil, just like grass."

Sorry to disappoint anybody, but hair and fingernails don't grow after death. Instead, our bodies dehydrate and our skin shrinks and tightens, pulling away from the hair and nails, creating the illusion of growth. Interestingly, funeral parlors put moisturizer on corpses to help reduce this effect.

Richard Siddall

It is quite a common error to believe that fingernails and hair continue growing after death. Some time ago a person convicted of murder asked my library's information service for literature relating to the effect. He wanted to prove his innocence by relying on the postmortem growth of hair and fingernails which would throw doubt on the timing of the killing. Unfortunately for the individual, no scientific verification for this growth exists.

Baerbel Schaefer
Marburg University Library, Germany

KEEPING PACE

If I was poisoned with a drug to stop my heart beating, would my pacemaker keep me alive?
Mark Rowney

It depends on the circumstances. Drugs affecting the heart's ability to contract act either on the specialized electrical conduction systems that trigger a heartbeat or on the ability of the heart muscle to generate contractile force. In the former case, a pacemaker will continue to stimulate cardiac contractions even when the heart's intrinsic electrical activity has been suppressed. In the latter, the drugs will prevent the heart muscle from contracting, so the pacemaker will not keep the heart beating. Cardioplegia, the technique used to paralyze the heart during open-heart surgery, uses the latter approach. Once beating ceases, surgeons can add bypass grafts or replace valves. When the cardioplegia is reversed, the heart muscle resumes its contractions.

Rafe Chamberlain-Webber
Consultant cardiologist

LAND OF NOD

I have difficulty sleeping at night and often find that it takes me up to two hours to nod off. Yet when I travel as a passenger in a car, bus, or train, I find that I can fall asleep in minutes, often unintentionally, and even while listening to music. I have roused myself from slumber at the end of a journey only to fall into bed and be unable to sleep. Why is it that the calm, relaxing, and comfortable environment of a bed is less conducive to sleep?
Alan Godfrey

I too used to suffer from the "fall asleep anywhere except in bed" syndrome. My conclusion was that it was mostly caused by my state of mind. When traveling, the noise and warmth can help drowsiness take over—coupled with the fact that one is usually not supposed to be doing anything. However, at bedtime it seemed that my mind would take advantage of the lack of external distractions to start some serious thinking. That, coupled with the anxiety of looking over at the

clock to see it creep past midnight, heralding another night of insufficient sleep, would contrive to keep me awake.

I have since found that distracting my mind from analytic, rational thinking helps a lot. Imagining mostly passive scenarios is particularly helpful. My own favorite is flying, usually just cruising along and admiring the view. After using this technique a few times, my anxiety also lessened and I can now frequently sleep easily without using this technique at all. Another tactic on particularly troublesome nights is to get up, move around a bit, and go back to bed.

I have also found that using my computer before bedtime is not conducive to falling asleep.

Richard Thomas

At last, someone who shares my problem. The reason I can't get to sleep is because I am so busy thinking, and I find it quite hard to tell my mind to turn off. I also find that after a long day, bedtime is the first time I'm not actually doing something, and this gives me time to think. Unfortunately, my mind will not turn off and the thoughts have to be allowed to run their course.

When I travel as a passenger in a car it is the opposite. I find that I sleep easily on long trips, when nobody is talking to me (usually because I am listening to music). The constant soothing vibrations of the car, the countryside flashing past, and exhaustion caused by bad sleep over many previous nights quickly send me to the land of Nod. The experience is doubtless quite similar to that of a baby being rocked to sleep, and I find that my brain doesn't think, it just processes the passing countryside.

The solution to all this, I have been told, is to do some strenuous exercise during the day. You will then be physically exhausted when you get to bed, and hopefully fall asleep straight away. This then means you will not be as tired when you are in the car, and consequently makes you less inclined to doze off.

If you are only concerned about falling asleep in the car, you could try to keep active and alert by opening a window or chatting with someone, although I prefer to catch up on some sleep while the going is easy.

James Ley

It is all too easy to associate bedtime with sleeplessness, often accompanied by the dread of yet another night's bad sleep. This means you are virtually preprogrammed to spend a couple of hours trying to get to sleep after lights out. Sleeping elsewhere helps break that link and allows more pleasant associations between bed and sleep to develop.

There's no point in going to bed if you can't sleep. Ignore the clock and don't worry about getting your seven or eight hours. The worst that can happen is you may be a little tired the next day, and this is not necessarily a bad thing, because you will probably sleep better that night. Be warned, though: if you are a poor night sleeper, a daytime nap lasting more than about fifteen minutes will only worsen sleep that night.

A car, like a train, produces very low frequency, inaudible humming known as infrasound, which many people find relaxing and soporific. These sounds come from the wheels running along roadways or railway tracks, as well as other resonances. That's why you find it easier to sleep while traveling than in your bed.

But what if you can't get to sleep at night after, say, twenty minutes? You should abandon the bedroom and do something absorbing in another room, such as putting together a jigsaw puzzle. It may sound ridiculous, but it's not as daft as languishing in bed. Searching for those myriad pieces of blue sky, for instance, is quite demanding and will take your mind off things, eventually making your eyes feel heavy—then it's back to bed.

If you do decide to get up in the night, there are a few things to avoid. Do not put on fluorescent or other bright lights—these make you too alert. So use a table lamp when doing your jigsaw puzzle. You should also avoid reading, watching TV, or listening to the radio because your mind can easily flip back to those worrying thoughts while immersed in an overstimulating drama.

Jim Horne
Sleep Research Center
Loughborough University, United Kingdom

Jim Horne is author of Sleepfaring: A Journey Through the Science of Sleep *(New York: Oxford University Press, 2006).—Ed.*

IFFY TUMMY

Tripe is the stomach lining of cows and other animals. So how is tripe digested by humans? You would assume that it would not be broken down by the normal digestive process. If it is, why is it an effective stomach lining?
Adil Hussain

Our digestive juices can break down tripe because it has lost the protection it had when it was part of a living animal. Inside our own digestive tract it is broken down by gastric acid, in the form of hydrochloric acid produced by parietal cells, and enzymes such as pepsin.

These substances could eat away at our own stomach lining too, were it not for the fact that it is covered in a layer of thick bicarbonate-rich mucus produced by so-called goblet cells. Bicarbonate is alkaline, so it neutralizes the acid secreted by the parietal cells. Without the continuous bicarbonate supply your stomach would digest itself.

When, occasionally, gastric acid does reach the stomach lining, it causes a gastric or peptic ulcer. It was once believed that the majority of ulcers were caused by gastric acid, but we now know that the bacterium *Helicobacter pylori* is more often to blame, and that drugs such as aspirin can make matters worse. *H. pylori* weakens the protective mucus coating of the stomach and duodenum, allowing acid to get through to the lining beneath. The bacterium is able to survive in stomach acid because it secretes enzymes that neutralize it.

Leigh Farina

Part of the answer lies in the fact that tripe is obtained from the ruminant fore-stomach, or reticulorumen, which does not secrete digestive juice in the way our own simple acid-secreting stomachs do. Cows eat grass and other fodder containing large quantities of cellulose. Mammalian enzymes cannot digest cellulose, so ruminants have evolved a complex stomach structure and a symbiotic arrangement to do the job.

The first three stomach chambers contain bacteria, protozoans, and fungi. These organisms can ferment plant cells to yield volatile fatty acids (VFAs), which are converted by the liver to sugars and other substances needed for energy,

growth, and lactation. The reticulorumen must be able to circulate and pulverize the ingested food before absorbing the VFAs, which is why it is a muscular organ with an epithelium that is both tough and able to allow VFAs to be absorbed. Tripe consists of this epithelium and the underlying supportive and muscle layers, making it a nutritious food.

That the "true" stomach is prone to self-digestion is evidenced by the formation of ulcers. Its only protection from this type of attack is an unbroken layer of tenacious mucus and trapped bicarbonate which is continuously secreted. After death, and with the mucus stripped away, an acid stomach would itself become highly digestible.

Ian Jeffcoate
University of Glasgow Veterinary School, United Kingdom

Stomach lining forms only one side of tripe. The other side is easily attacked and digested. Chewing also disrupts tripe's structure, leaving it open to enzymatic attack. Because it is dead tissue it has no effective repair processes. In addition, as with many structures that are resistant and self-repairing in a living organism, the process of cooking destroys tripe's integrity.

Richard Lucas

CAUGHT RED-BLOODED

If I committed a crime the day after having a full blood transfusion and left some of my blood at the scene of the crime, would forensic scientists be able to detect my DNA in it? Or would their analysis result in confusion? I hasten to add that

it is unlikely that I would feel well enough to commit any crime after a transfusion, and I also have no intention of trying this out.
Mark Blackmore

Blood has three cellular components: red blood cells, white blood cells, and platelets. Only the white cells are complete cells with nuclei and DNA. These three elements are typically separated when donated blood is processed. The blood used for transfusion only contains red cells, which lost their mitochondrial and nuclear DNA during development.

Even a complete exchange transfusion, where the recipient's blood is withdrawn as new blood is added, would not result in the donor's DNA being left at a crime scene. Large numbers of white cells are located outside the bloodstream. On the day after a transfusion, the recipient's blood would pick up a considerable number of these cells. So the recipient could still be identified.

Whole blood is rarely used in transfusions. When it is, the white cells are attacked by the recipient's immune system, making it highly unlikely that any of the donor's DNA would remain the following day.

The only exception to this is a bone marrow transplant. The stem cells in the transplant populate the recipient's blood with the donor's DNA and, because their own bone marrow has failed, little or none of their DNA remains. This might indeed confuse the evidence at a hypothetical crime scene.

There is a simpler way for master criminals to throw the police off the scent. If they have blood samples from other people, these could be placed at the scene. Or anonymous donor red blood cells could be mixed with DNA amplified from a hair from someone they want to frame to create a bloodlike residue.

Picking up an ashtray of cigarette stubs from a public place and leaving these at a scene would be the simplest way to create confusing DNA evidence.

Ian Flitcroft
Mater Misericordiae Hospital, Ireland

A full blood transfusion, in which 100 percent of the patient's blood is replaced, is not possible. Even an exchange transfusion, like that given to some sick newborn babies, can only replace 60 percent of the patient's blood. So if the questioner had an exchange transfusion and left a bloodstain, it would still contain some of his DNA.

A more interesting question is whether such a bloodstain would also contain the donor's DNA soon after a transfusion. Probably not. True, after receiving a transfusion some of the blood in a bloodstain will be from the donor, but red blood cells do not have a nucleus.

White blood cells, which do contain DNA, don't survive when cooled to 39.2°F (4°C), the standard storage temperature for red cells. So the relatively few white blood cells present in a transfusion are almost all dead by the time the blood is given. These remnants, and any miraculously surviving white cells, are quickly recognized as foreign by the recipient and are eliminated.

The day after a transfusion there won't be any circulating donor white cells left, either dead or alive.

Nicolas Slater
Consultant hematologist

TUMMY CHILLS

When I wade into the sea or immerse myself in an outdoor swimming pool, why does the water always seem coldest when it reaches my midriff?
Marion Hurden

When we feel hot or cold, what we actually perceive is the difference in temperature between ourselves and another object.

Take three buckets and fill one with freezing cold water, one with lukewarm water, and one with hot (not scalding) water. Then simultaneously plunge your left hand into the cold-water bucket and your right into the hot-water bucket and leave them there for thirty seconds before removing both of your hands and putting them into the third bucket containing the lukewarm water. Your cold left hand will feel that the lukewarm water is hot and your hot right hand will feel that the same lukewarm water is cold.

So in the original scenario the reader is really describing a greater temperature difference between tummy and sea than between legs and sea. Therefore the question now becomes, "Why is a person's tummy area warmer than a person's legs?"

I would postulate that the difference comes from heat generated by the stomach and intestines as they digest meals, a process which continues for a long time after food is eaten. This would give a localized increase in skin temperature in the tummy area and cause the sea to feel coldest when it reaches that level.

Oonagh Griffith

Two factors operate when we enter the sea. First, our skin temperature is cooler than our core temperature—that of our trunk organs, heart, liver, and stomach. This is because we circulate warm blood through cool skin to lose body heat.

Second, our limbs are typically cooler than our trunk because we operate a countercurrent heat exchanger to limit the loss of body heat. The arteries that supply a limb run close to the veins returning from this limb. Warm arterial blood flowing to the limbs is cooled by venous blood returning to the heart. This means that our feet and hands are almost always cooler than the skin of our trunk.

Both of these factors mean that the difference between skin temperature and water temperature becomes greater as you immerse yourself. Wearing a bathing suit makes the midriff even warmer than it would be if you were a nudist. The bigger the difference, the greater the discomfort.

Peter Bursztyn

Several of our male correspondents point out that the shock is worst when cold water reaches their testicles. This is odd because these are cooler than the abdomen, with less of a temperature difference between them and the sea. Sexual organs are, of course, richly supplied with sensory nerves, so it would seem as though temperature difference is not the whole story.—Ed.

WASTE DISPOSAL

Excreta from babies that are fed on breast milk alone tend to be fairly odorless but if the babies are switched to infant formula

milk you get the infamous baby poo smell. What ingredient in infant milk formula causes this, er, interesting phenomenon?
Tolu Akinola

The protein (mostly lactalbumin) and fat in human breast milk are more easily digested by babies than the protein (mostly casein) in cow's-milk formula, which presumably makes the end product less smelly. However, breast milk also has a laxative effect, so breastfed babies are often notoriously prolific when it comes to filling their diapers—an interesting trade-off.

If you think the diapers of formula-fed babies are bad, wait until you have a meat-eating toddler. My husband and I play elaborate games in order to avoid changing our daughter's diaper—pretending not to smell it, having urgent chores to do, faking sleep, and suchlike.

Compared with a toddler, a baby diaper smells lovely, regardless of whether they are fed breast milk or formula.

Rebecca Rose

As a new mom I have also asked myself this question. I breastfed my daughter, and she had diapers of the none-too-stinky kind, whereas moms in my playgroup who formula-fed their children complained about the "stinkiness" of the diapers that they changed.

Here are my findings, most of it insight from my mother-in-law, who is a lactation consultant.

Human babies have evolved to drink human milk for optimum efficient nutrient absorption, and breast milk is tailored to the baby's age, changing constantly throughout the hours, days, and months of an infant's first year.

Most formula is cow-milk based, and therefore differs

from human breast milk in many ways. It contains much more protein of many differing types, fats, and elements such as aluminum, manganese, cadmium, and iron.

Whereas breast milk is, in effect, made to order for the individual baby and its age, formula has to conform to an average by suiting a variety of infants of differing ages. So, for example, a baby of one week gets the same nutrients as a child of twelve months. Because of this, much of the fat and protein will be excreted instead of being absorbed by the body. Another factor is the excess of iron in formula (one reason why formula-fed babies are often constipated). The iron in formula is poorly absorbed, so a large amount is required, most of which is also excreted.

Breastfed babies absorb almost 100 percent of what they take in—and rarely become constipated. Their excreta are mostly water, with very little protein, fat, or trace elements. The excess fats, proteins, and nutrients of formula milk that a baby cannot absorb lead to the stinkier feces in non-breastfed babies.

Jo Resnick

DIMINISHING RETURNS

Today I walked up a hill that, in the past, has had me in a sweat and out of breath. Now, a lot fitter and a little lighter, it took less effort. So given two people of the same weight but different levels of fitness walking up the same hill at the same pace, does the fitter person burn fewer calories?

Eleanor Stanton

You cannot escape thermodynamics, even for the human musculoskeletal and circulatory systems. At a given rate of work, whether level walking or running, or climbing a hill, the total amount of energy used by individuals of the same weight will be the same, so long as they are of similar mechanical and metabolic efficiency (more on that later). Identical absolute amounts of work require the same amount of energy, and thus fit and unfit people doing the same task use the same amount of calories (or joules, in the international system, or SI, units)—again, as long as they have the same mechanical and metabolic efficiency.

This is hard to prove for walking and running because style is important. Fitter people often adopt an economical style of movement that increases efficiency to a small extent: world champions rarely run with swaying hips, rolling heads, and flailing arms (though some manage to win using very strange styles, such as Michael Johnson, the world-record-breaking two-hundred-meter runner who looked as though he was almost leaning backward). However, when careful studies are carried out on trained and untrained persons working on bicycle ergometers, the work efficiency is identical (Martin Mogensen et al., "Cycling Efficiency in Humans Is Related to Low UCP3 Content . . . ," *Journal of Physiology*, vol. 571, no. 3 (2006), p. 669).

This suggests that the biomechanical, cardiovascular, and metabolic adaptations shown with increasing exposure to physical work—in other words, training to become fit for the task—increase capacity without increasing efficiency. This is not to say that efficiency is not important to physical performance, but the underlying characteristics are likely to

be mainly genetically determined, rather than acquired by training.

In the paper mentioned above, Martin Mogensen of the Institute of Sports Science and Clinical Biomechanics, University of Southern Denmark, Odense, and his colleagues showed the most efficient subjects—trained or untrained—had a greater proportion of a kind of muscle fiber called fast-twitch type 2 in their legs. In addition, the world champion East African distance runners seem to have a mechanical advantage in that their calf muscles are placed nearer to their knees, lessening the work needed to be done by the thigh muscles to swing the lower leg. These factors, which increase efficiency in individuals, are mainly determined by our genes. It may even be that the ability to improve capacity by training is itself to some extent genetically determined.

Michael Rennie
Professor of Clinical Physiology
University of Nottingham, United Kingdom

Moving two equal masses by a given distance uphill will require the same energy expenditure for both masses, but the scenario is more complex than is implied by this simple analysis. At a cellular level, the conversion of nutritional substrates into mechanical energy by skeletal muscle has an efficiency of about 30 percent and is largely independent of one's level of fitness during normal aerobic metabolism.

However, the less fit person will have a lower anaerobic threshold (AT), which may be exceeded as they work to climb the hill. Beyond this point, the efficiency of energy extraction from substrates will steadily fall for the unfit person, so their total caloric consumption will, in this case, increase.

Furthermore, even if body weights were identical, a greater proportion of the less fit person's body will be fat, and some energy will be expended in the pendular motion of this deadweight in the legs as well as other parts of the body. The questioner will therefore perceive some benefit because of her noted weight reduction.

Regularly climbing the hill constitutes a form of training, which will have increased the writer's AT such that the AT may now be higher than the workload required to climb the hill at the usual pace. Workload that is below one's AT is well tolerated for extended periods, whereas that beyond the AT is unsustainable and accompanied by a sensation of distress that will be unpleasantly familiar to anyone who has undertaken strenuous exercise.

Mark Colson

ETERNAL YOUTH CLUB

In the sixteenth century, so the story goes, the Hungarian countess Elizabeth Báthory bathed in the blood of young girls in a bid for eternal youth. More recently we have learned that the telomeres of our chromosomes become shorter as we get older, and this seems to be related to aging. Without wishing to condone Báthory's deplorable sadism, if one were to take a blood sample from an infant, store it perfectly for fifty years, then reintroduce it to the body of the adult, could it have any positive effect?

Barbara Robson

There are really two separate questions here. The first is, what are the causes of aging? Telomere shortening is one theory but it can't really explain aging because many animals such as the nematode worm age and die without undergoing cell division at all. Conversely, cancerous cells can effectively be immortal, undergoing thousands of cycles of cell division without any reduction of potency. Aging is a complex interplay of many different phenomena including a gradual decrease in mitochondrial function because of oxidative stress and the buildup of misshapen proteins resulting from transcription mistakes and accumulated DNA damage.

The second question is whether a transfusion would work. The answer is no. Replenishing "aged" blood with "young" blood would not ameliorate any of the cellular phenomena that lead to aging. The most probable outcome would be negative: the person involved would quickly become sick after the transfusion because the replaced blood would lack the circulating antibodies that the individual had built up over the preceding fifty years. As a result, germs that had not been a problem prior to the transfusion would suddenly find a new and easy target in the new blood circulating through the body.

Allan Lees
Chief Information Officer
Buck Institute for Age Research, California

Even if we knew what role the shortening of telomeres plays in aging, telomere transfusions could hardly help. Telomeres are repetitive sequences of base pairs that act as disposable buffers at the end of chromosomes. In somatic cells (the general cells, not including reproductive cells or new stem cells, that make up our bodies) the telomeres undergo shortening during division, shortening that in reproductive cells would cost genetically useful material.

Cells with long telomeres do nothing to protect other cells that have lost their own telomeres. Each telomere affects only its own end of its own chromosome in its own cell.

In reproductive cells such as oocytes and spermatogonia, a special enzyme called telomerase extends the telomeres to a good starting length. The process continues at least to the early stages of embryonic development and persists in some classes of stem cells. In particular, most blood cells are short-lived, so they have to be continually replenished from stem cells in the spleen, marrow, and so on. This means a frozen autograft from these structures might serve to replace the stem cells of some critical tissues late in life, but blood transfusion would not. Stimulating telomerase production might work better, but it also could be risky because this is how some kinds of cancers survive.

Frank Horseman

SWEAT ON THE HOOF

Why, after going on a long run, do I only start to sweat profusely immediately after I have finished? Other runners I have spoken to also experience this.

Milan Harbin

When sweat evaporates it creates a layer of saturated air close to the skin, which inhibits further evaporation. During a run, the movement of air relative to your body replaces this with fresh air, allowing evaporation to occur.

Every gram of sweat that evaporates takes 2,260 joules of body heat with it. When you stop running, the layer of saturated air builds up and the sweat does not evaporate. That creates the perception that you have suddenly started sweating profusely, and also raises skin temperature—which does, in fact, make you sweat slightly more. Proof comes from the fact that the phenomenon will not occur when you stop running on a windy day.

Shane Maloney
School of Biomedical and Chemical Science
University of Western Australia

You actually begin to sweat not long after you have started to run, once your muscles have settled into a working routine and you are using more energy than you would if you were resting. What you are experiencing is the windchill effect of your own movement: as a result, the movement of air over your skin and through your clothes wicks the moisture away from your body before it can build up. Try wearing a small backpack or taping a section of something like plastic cooking wrap over an area of your chest. Sweat will build up here and not elsewhere, even after quite a short period of exercise.

When you stop running your muscles still have reserve heat to expel and this, combined with less air movement over your body, means that sweat will build up on the skin. If you warm down after a run, rather than stopping suddenly, the buildup of sweat may well not occur.

Relative humidity and ambient temperature also have a huge effect on how much sweat is produced. Higher temperatures plus high humidity leave you more sweaty than cooler and less humid conditions. Try running through a forest on a sunny day after rain compared with open country in a breeze on a cooler day.

Dave Banks

There is a second, minor effect beyond evaporative cooling at work here. When you stop exercising, your muscles stop working and generating heat, but your body's "thermostat" is still set to a higher temperature. As it gradually resets to a resting state, you continue to produce sweat to cool yourself down.

David Gibson

CRUEL WORLD

Scientists have worked out an evolutionary basis for many behaviors, such as altruism and jealousy. Over the millennia, however, people have been unbelievably and gratuitously cruel to each other. The evolutionary advantage of cruelty is not obvious. What is its biological basis in humans?

Brian Kavanagh

There may not be an evolutionary advantage, at least not any longer. It is more likely a throwback from the past. Not so very long ago in evolutionary terms, humans lived as hunter-gatherers in small groups. Indeed, many such groups still remain, in the remnants of tropical rain forests, for example.

There is safety in numbers in such communities and individuals can specialize in what they are good at, knowing that other essential tasks will be taken care of by their companions. However, although in-group individuals may be

loving and caring toward each other, there is open hostility toward any out-group that competes for territory and food resources. Neanderthals (*Homo neanderthalensis*) were one group that lost out to humans (*Homo sapiens*) in such past conflicts. However, to speak of cruelty in an evolutionary context seems inappropriately anthropomorphic, despite the fact that we are discussing human beings.

Homo sapiens evolved in parallel with the other great apes, in which we can see similar "cruel" behaviors. Even within their own groups, great apes are not necessarily loving and caring. They nevertheless gain advantage from living in a group, through increased security and the sharing of tasks. Such hunter-gatherers have to carve out and defend a territory for themselves in order to survive. In consequence, outside groups are a threat to their existence. A group of apes has its hierarchy, and so a conflict between two groups is a threat not only to survival, but to the dominance of the males in a particular group. The conflict is resolved by either driving off the rival group or by killing all the rival males in the group and assimilating the females and young. Whichever group wins the conflict is obviously the fittest in terms of survival. That's what it's all about in the end, cruel or not.

Modern humans behave in a similar fashion. It is quite clear even today that humans belonging to one group can easily be influenced to see humans of another group as being subhuman and inferior, enabling their extermination without mercy. This is true even when the differences are over beliefs, rather than over limited resources.

The Americas were recently colonized by Europeans, who pushed the Native Americans aside and killed any who resisted. Hitler's motivation was just the same in the Second World

War. He envisaged a greater German territory, stretching to the Urals, into which the "superior" Aryan race could grow. The process repeats itself regularly, most recently in the Balkans and Darfur.

Terence Hollingworth

Only one word is needed to answer this question—power. Let's leave women out of it. Very few women, if any, have reached the heights of cruelty of Nero or Caligula—or Saddam Hussein for that matter. The would-be alpha male must trample over all opposition, must create fear to keep himself in power and must keep enough of his cronies loyal. He thus obtains access to a great number of women, the best in material goods and food, and assures himself of the largest possible number of healthy descendants. Other methods of the cruel alpha male may include reducing the number of offspring of rivals, who might compete with his own children.

Valerie Moyses

Your previous correspondent seeks to exclude women from the discussion of the survival of unbelievable and gratuitous cruelty as a human trait. They have not been so dominant in the history of large and dramatic behaviors, perhaps because they have seldom been in a position to initiate them. When they are in such a situation we see such exceptions as the sixteenth-century Hungarian serial killer Elizabeth Báthory (see "Eternal Youth Club," page 75) and ancient Roman females of the family that produced her role models, Caligula and Nero.

Today, in the everyday world, wannabe alpha females can create just as much nastiness as males, if not more. As a

primary schoolteacher I have to deal with bullying. In forty years, boy bullies, as opposed to boys who sometimes bully, have been rare. Female bullying is much more common, much more difficult to deal with, and delivers much more long-lasting, though less visible, damage. "You can be my friend if you promise never to speak to her ever again" is the style of approach used by girls.

I did ask the hangers-on once why they wanted to be the friend of someone who had demanded they dump another friend. They had no answer, even expressing disbelief that I should ask, despite being upset by the situation. Girls bullied by other girls are excluded, made to feel of no value, and have been driven to suicide. Females cannot be excused complicity in the production of cruelty.

I don't know why the trait of cruelty survives, but I do know why the trait of compliance survives. Survival depends on not being cast out from the central group. And while that trait persists, so do the queen bees.

Penelope Stanford

UNDERGROUND RESERVOIR

Is there somewhere in the body that can store fluid, apart from the bladder? Most nights I wake to empty my bladder after exactly four hours' sleep, but on occasional nights where I haven't been to the toilet all day previously I have to get up three or four times instead of only once. It's obvious that the urine has been building up during the day, but it wasn't there in my bladder before I went to bed. So where was it?

Virginia Love

Is there somewhere in the body that can store fluid, apart from the bladder? There is, and it's simple when you think about it—it's the bloodstream. Faced with a bladder which is full and hasn't been emptied all day, the kidneys cut down on urine production. When the bladder is emptied at last, they work to restore the blood's correct electrolyte balance, and the bladder soon fills again.

Kidney activity is to some extent also controlled by the body's clock, and less urine is produced at night. Most people urinate several times during the day but will not have to get up during their sleep. If you reverse the cycle by flying halfway round the world, you find yourself not needing to pee all day then getting up all night. This is undoubtedly a trained response and your questioner seems to have trained her body into the opposite cycle—making more urine at night—which may suit her lifestyle or job.

Guy Cox
Cell biologist
University of Sydney, Australia

One answer is that the fluid is found in your legs and feet. Patients with any degree of heart failure or kidney disease know this well. Fluid pools in the legs during the day, and as soon as the body is horizontal the fluid reenters circulation, straight through the kidneys and into the bladder.

You don't have to be sick to see this. If you've ever been on a long flight, consider why you always have to find a toilet while standing in the queue for customs. Prolonged inactivity causes extracellular fluid to pool in your lower half. This is why it is hard to tie the laces on your shoes when it's time to disembark. Once your leg muscles start pumping

the extra fluid back into the general pool, you soon need to go to the toilet.

Michael Carrette

A 150-pound man is 65 percent water. That amounts to approximately twelve gallons (forty-five liters). Most of this is contained inside the trillions of body cells. A lesser amount surrounds these cells, while an even smaller amount circulates as plasma. Fluctuations in body weight over hours, days, or a month are a feature of the body's ability to vary this volume. Variable tightness of finger rings, end-of-the-day aching legs, and tight shoes are testimony to this movement of fluid.

To answer the question—the water is everywhere in the body and there is a lot of it. Urine is just the fluid kept in the bladder as a result of renal activity. The need to urinate actually results from only two hundred to three hundred milliliters—less than one-tenth of a teaspoon—of urine in the bladder.

Consuming a lot of salt encourages retention of fluid and can increase blood pressure, while several complex hormones, such as antidiuretic hormone and aldosterone, play a significant part in this process of fluid balance. So does estrogen, leading to weight gain during some parts of the menstrual cycle.

Renal function slows during the course of the night, allowing most of us (but not the author of this question) to get a good night's sleep. Bladder tone also decreases at this time, enabling it to hold more fluid.

John Curran

WHAT'S THAT CRACK?

I regret to say that I have a habit of cracking my knuckles. I've read somewhere that it does no harm, but I am still far from convinced. Could it be damaging in the long run? And if so, why?

Alex Cowley

There are a couple of theories regarding what causes the "pop" heard in knuckle cracking. The common view is that it is caused by bubbles of gas within the synovial capsule of the metacarpophalangeal and phalangeal joints. The bubbles form when these joints are stressed—a process called cavitation—and then collapse as the pressure changes within the joint. It is the collapse that creates the noise. The energy released by this process has been estimated at just 0.07 millijoules per cubic millimeter. To cause damage to a joint, this figure would have to rise to about 1.0 millijoule per cubic millimeter. Cumulative damage from these pops cannot be ruled out, however.

The second theory is that the noise comes from the sudden deformation of the fibrous joint capsule itself and that the pop is its sudden slap onto the joint fluid within. This might cause microtrauma, which could accumulate over years.

Regardless of the theories, there is little evidence that knuckle cracking causes arthritis: a survey of knuckle crackers showed no more incidence of arthritis than non–knuckle crackers. One American doctor went so far as to crack the knuckles on just one hand for fifty years to see if there was a difference between that hand and the other—there wasn't.

It is possible to cause acute trauma from the stress required to cause the joints to pop in the first place, of course, but one has to say that it's quite satisfying, isn't it?

For more on knuckle cracking, visit http://student.bmj.com/back_issues/1201/life/477.html.

David Farnsworth

The sound is caused by a bubble of nitrogen gas forming in the joint. This occurs through a pressure drop that is created when the joint is forced to the extreme of its range of motion. After the "pop," the joint capsule is temporarily enlarged, which also increases the neural firing in the joint's proprioception receptors. These signals transmit over the local nerve root. This increased neural activity inhibits pain signals from smaller nerve fibers in the same dermatome—the area of skin supplied by the same nerve root.

The only known effect of repeated joint-popping is in the joints of the spine, and results in a reduction in the financial content of the patient's purse: chiropractic spinal treatments rely on eliciting the sound—and offering brief relief—but later when symptoms recur (presumably as the gas is reabsorbed) the cyclic need for treatment resumes.

Don L. Jewett, M.D.
Professor Emeritus of Orthopedic Surgery
University of California–San Francisco

We expect some chiropractors will disagree with this opinion. Our Web address is in the front of the book.—Ed.

COUNT THE RINGS

My partner's father was born in Cameroon before proper records were kept. He would like to know how old he really is. Is there any way to determine the exact age of a living human being?
Peter White

Age evaluation is a growing area of involvement for forensic practitioners: it is often needed for individuals seeking asylum or refugee status, who may be genuinely unaware of their age. A requirement is placed on the practitioner to assign age with the greatest degree of accuracy possible, because there can be big benefits in being assigned a certain age.

The relationship between actual age and "biological age" is strongest in the young. We can assign the age of a child to within a few years, but even here it is not an exact science, because environmental and genetic factors will have an influence on how the body ages. Age evaluation is much more difficult in the older generation, with a larger margin of error.

All evaluations should start with a psychological assessment to identify information that the individual can remember or perhaps incidents that they were involved in. These can help the clinical assessor to home in on an age bracket for the individual. Beyond this, we must rely on skeletal indicators. Dental age is useful in gauging a juvenile's age, but of limited help for older people. Assessment of the skeleton requires the use of X-rays (although in some cases this raises ethical problems). Either flat-plate radiography or CT (computerized tomography) images are ideal.

In older people a combination of factors helps to build

up an age picture. These might include the degree of closure of the cranial sutures; evidence of degenerative conditions including osteoarthritis and the degree of ossification of cartilaginous structures, such as the costal cartilages that give the thorax its elasticity; changes to the laryngeal apparatus and the pubic symphysis; and even the extent of bone loss.

There is no single feature that will assign age with accuracy in an adult, but a multifactorial approach, combined with an understanding of the variations between populations, can provide a good estimate.

Sue Black
Center for Anatomy and Human Identification, College of Life Sciences
University of Dundee, United Kingdom

It is possible to determine the age of a living human with a considerable degree of accuracy up to the age of sixty or so, beyond which it becomes more approximate.

Tables are available detailing the time of eruption of the permanent and temporary teeth. As an example, the first permanent molars erupt at around six years of age, with second molars at twelve, while the presence of four wisdom teeth usually means the subject is at least eighteen years old. Gauging the calcification of their roots may also help to arrive at a rough estimate of age. While tooth eruption can be externally and visually judged, observing calcification requires an X-ray. If the roots of all teeth are calcified, the person would be older than twenty-five.

There are two other dental methods. One is Boyde's incremental lines—using the striae in tooth enamel, which change with age—and the second is Gustafson's method—which assesses six age-associated parameters of teeth. Both of these are useful for judging age up to sixty years.

An X-ray of the wrist in children, of the elbows and knees in adults, or of the skull and the vertebrae in older people can help to deduce age. That's because precise tables of age-related ossification—the hardening of the bone ends—are available which incorporate different races, dietetic, and geographical factors. X-rays of the shoulder, hip, ankle, and pelvis help to arrive at an even closer figure.

Between the ages of twenty-five and sixty, the meeting points of the skull bones, the larynx, and the intervertebral discs can be studied. All this in conjunction with height and weight, which are thought to be the least reliable factors, gives a good indication of age.

Above the age of sixty, examining eyes for cataracts and corneal opacities and checking for graying hair can be useful, although beyond sixty years it is very difficult to arrive at a precise age.

Vivek Jain

DEXTROUS DILEMMA

Why are some people left-handed and others right-handed?
Leila Gabasova (age twelve)

The simple answer is that they've inherited genes for left- or right-handedness, which is why handedness runs in families and identical twins are more likely to have the same handedness than dizygotic (fraternal) twins. The genes involved are a little strange, because while one makes people right-handed,

the other only makes it random as to whether an individual is right- or left-handed. So identical twins with the latter gene can have different handedness.

Genes are only the immediate cause of handedness, but very occasionally, "biological noise" during development, or brain or arm trauma, will override genes and cause "pathological handedness."

Why humans alone among animals are 90 percent right-handed is a separate question, with the answer going back two million years. This is when human brains became asymmetric and the neural equipment for the fast, precise movements for speech and finger dexterity became localized in the left hemisphere. Why it is the left hemisphere is unclear.

Yet another question is why some people are left-handed. The answer is that there must be advantages to having the genes for left-handedness, although the advantages are still to be discovered. Finally, why aren't all animals ambidextrous? Most likely because it pays to specialize—if all practice is with just one hand, that hand will be more accomplished than a hand that only benefited from half the practice time.

Chris McManus
Professor of Psychology and Medical Education
University College London

Chris McManus is the author of Right Hand, Left Hand *(Cambridge, Mass.: Harvard University Press, 2002).—Ed.*

Your earlier correspondent was not entirely correct when he stated that there must be advantages to having the genes for left-handedness. Surely it is only necessary that those genes, in any given environment, confer no relative disadvantage?

Kevin Donaldson

The previous correspondent is correct. And in fact he doesn't go far enough: even if left-handedness is a disadvantage, there are all sorts of reasons for genetic variations favoring left-handedness to persist. The variations might protect against a disease, or provide some other advantage unrelated to left-handedness, or merely sit next to another genetic variation that is highly advantageous. In addition, new mutations favoring left-handedness could be arising faster than natural selection can eliminate them from populations (evolution is very weak in humans). It could also be related to the number of gene copies, rather than a specific variant, and so on.

Alex Bentley

MIDDLE-AGE SPREAD

Why, as we age, do we find it easier to gain body fat and, of equal importance, why is it so much more difficult to get rid of it?

Andy Gravelius

There are a number of contributing factors at work here. As we age we tend to reduce our active pursuits and, above the age of twenty, our resting metabolic rate reduces by about 3 percent per decade, allowing our bodies to go further and further on the same amount of food. On top of this, as we age, we use less energy to digest and process the food that we eat.

There are also hormonal changes that encourage fat buildup. A decrease in levels of growth hormone and testosterone contributes to an increase in fat mass. Additionally there is a reduced response to hormones produced in the thyroid gland, which stimulate metabolism of lipids and carbohydrates, and possibly also to leptin, which helps to control appetite.

One more factor is our gradual reduction in height and a reduction in the weight of organs and muscle mass. For example, neurologists Anatole Dekaban and Doris Sadowsky identified a progressive fall in brain weight from about forty-five years of age onward.

Most of us do not compensate for these factors by reducing energy intake, so we gain weight. An article written in 2005 and explaining this in greater detail can be found in the *American Journal of Clinical Nutrition* (Dennis T. Villareal et al., "Obesity in Older Adults," vol. 82, no. 5, pp. 923–25).

Given this combination of factors, it is easy to understand why we find it more difficult to get rid of our excess weight as we age: our bodies are conspiring against us.

However, if we reduce our total energy intake while increasing our activity—especially activity that builds muscle mass—we have a better chance of losing, or at least maintaining, weight.

Cathy Watson

Many physiological details of our adipose tissue, where fat in the body is stored, are poorly understood. However, adipose tissue varies greatly, so some people have more troublesome spare tires than others.

One reason is that in our recent past there has been evolutionary selection for different patterns of accumulation of mature fat. In frigid climates, bears and humans need layers of blubber, which in the tropics could be fatal. In contrast, aboriginal peoples in hot climates with intermittent food supplies tend to carry fat on the belly, buttocks, or outer thighs, just as camels store fat on their backs.

There are functional differences between the fat of human babies and adults. Babies' "brown fat," a specialized form of adipose tissue, counters hypothermia, and youngsters' puppy fat is readily mobilized for growth and activity. However, adults' love handles are for reproduction, hard times, and famine, and to deplete strategically deposited stores too readily would be foolish.

So remember, fat is there for a purpose and as such should not be dismissed unappreciatively. It must manage fuel storage, conversion, and mobilization; it has elaborate endocrine functions, both metabolic and reproductive; and it differs drastically between populations as well as between people. Really, to abandon such a marvelous organ to either anorexia or obesity is obscene.

Jon Richfield

If you accept that the existence of menopause in women has an evolutionary advantage, by ensuring that a grandmother has time to care for her grandchildren, then it is not hard to imagine that grandparents need fat to tide them over when food is scarce and they need a reserve of energy.

This is because, in ancient societies, old people would be the last in a family group to eat, as the food supply was controlled by the stronger parents. If this argument from evolution is correct, it would explain why the metabolic mechanisms of the body become so efficient at storing fat as we get older, and in turn why it is so difficult to oppose these changes.

The same evolutionary argument applies to patterns of

sleep and wakefulness in the elderly. They nap in the day-time when others are on guard, but are awake at night, to keep fires going or be alert for predators.

Don L. Jewett, M.D.
University of California–San Francisco

4 FEELING OK?

BLOW IT OUT

Right now I have a cold. After blowing my nose for what seems like the millionth time, I wondered just how much mucus the nose produces during the average cold, and does its loss mean I lose any substantial amount of weight?
Nick Brown

On average the normal nose produces about eight ounces (a cupful) of mucus every day. During a cold there is additional flow. Most of the mucus produced normally flows down the throat and gets recycled within the body. During infection the nasal passage becomes constricted and therefore the inward passage of mucus is obstructed and it flows out through the nostrils.

There are other factors that increase the mucus flow, like the excessive formation of tears, which can make their way to the nasal passage and mix with the mucus. Normal body processes lead to the formation of these liquids, and their loss or gain is compensated for by absorption or excretion of water.

Saifuddin Ahmad

Speaking of "the average cold" means little because colds vary so greatly with virus or victim. The most important cold viruses, when checked by your immune system, settle permanently in your cells. Afterward, a bout of poor health or an infection such as flu may unchain old colds to flare up in the guise of new diseases. The resulting variation in texture and volume of snot is amazing.

What little liquid a healthy nasal cavity does produce—with no tears flowing, and no whiff of onion or allergen—usually passes back down the pharynx for swallowing. Over the course of a day a mild cold might produce a few milliliters of snot for blowing. Heavier colds, demanding the handkerchief twenty times an hour at say two to ten milliliters (less than two teaspoons) per blow, could cost you as much as 6.5 ounces an hour, and you then must drink liquids to make up the volume.

Nasal flow usually slows at night, but really serious secretion can force one to sleep sitting up to avoid swallowing phlegm and saliva.

Jon Richfield

DIRTY MONEY?

Can viruses and bacteria be transmitted on coins or notes, which pass through so many hands? What is the likelihood of catching something unpleasant from money, and if it does harbor disease, what might I catch? Finally, what is the average number of germs likely to be hitching a ride on coins or paper money?

Michaela Lanzarotti

Transmission of microorganisms is possible from any place where they are attached. How many will be transferred from coins or notes depends on a series of factors such as the number of organisms present and their ability to survive in such a dry environment, in which many will die.

The form of contact also makes a difference, whether it be by touching contaminated money, which can transfer the organisms to the hand, or the "hoovering" that occurs when people snort drugs through a banknote. The latter is obviously a more direct way of carrying germs into your nasal passages.

The number of individual viruses or bacteria needed to make you ill also counts. For viruses, it can be as high as a hundred thousand or as low as ten. One more factor is whether any germs taken in from money will have easy access to the site in the body where they can thrive.

In a recent microbiological investigation of money in the Netherlands it was observed that coins may harbor up to a thousand bacteria, and paper money a few million. Of course this depends on the type of material the money is made of: coins usually carry few microorganisms, which can also be the case for some materials used in producing paper money. We have very little reliable data about which viruses are present, although results on that topic may be available soon.

Rijkelt Beumer
Department of Food Technology and Nutrition
Wageningen University, The Netherlands

Viruses can survive on banknotes, as scientists in Switzerland recently showed when they dripped flu virus onto notes accompanied by human nasal mucus. The viruses remained

viable for up to seventeen days (see http://www.newscientist .com/channel/health/dn12116).—Ed.

BITING BACK

Why do mosquitoes bite one individual but not another? I am traveling around Asia with my girlfriend. I don't appear to have been bitten at all, while she has been bitten endlessly. We sleep in the same bed and share a similar diet and neither of us wears scented products.

Matthew Hedley

Are you sure that you don't just fail to react noticeably to mosquito bites? Some individuals hardly react at all to bites, which interferes with reliable bite-scoring.

Attractiveness to mosquitoes correlates with obesity, maleness, temperature, and sweat, but personal differences in sweat composition and skin bacteria have unpredictable effects. Mosquitoes are said to favor certain blood groups, but there is more argument than hard evidence on which groups they prefer, if any, or on whether diet makes any difference.

The most definite attractants are lactic acid, carbon dioxide, warmth, certain fatty acids, and other compounds from sweat or its bacterial decomposition. Notoriously, Limburger cheese and unwashed socks are rich sources of such chemicals.

Some cosmetics are effective repellents or, more accurately, they counteract attractants by confusing the mosquitoes' senses.

Avram Sherwin

While it is true we all produce a bouquet of natural chemicals, some of which attract biting insects, a fortunate few give off an aroma that apparently masks these attractive chemicals, and so prevents these people from being bitten because the mosquitoes can't track them down. James Logan and John Pickett at Rothamsted Research in Hertfordshire, United Kingdom, have pinpointed the human chemicals that keep bloodsucking insects at bay and are developing a mosquito repellent based on these chemicals.

The team at Rothamsted first noticed that some individuals are less appealing to bloodsucking insects when they saw some herds of cattle having to fend off flies while others grazed undisturbed. Pickett's group reasoned that animals in the undisturbed herd must simply smell less alluring to the flies and examined the cows' chemical profiles. Sure enough, they discovered that very distinctive chemical signals emanated from certain individuals, and these lucky cows were unattractive to flies.

Logan discovered that humans have distinct chemical signatures too. He let yellow fever mosquitoes (*Aedes aegypti*) fly along a Y-shaped maze and wafted the scent from volunteers' hands down the prongs of the maze. While some chemicals attracted mosquitoes, others repelled them, and so the insects flew down the more attractive prong of the Y. Logan has since been able to identify which chemicals the insects respond to by strapping miniature electrodes to the antennae of female mosquitoes and checking their responses. It turns out that only some humans produce the masking chemicals that mosquitoes find unappealing.

Yfke van Bergen

A POX ON YOU

Why does removing chickenpox scabs before they heal leave a scar, while removing normal scabs does not?
Thomas Oxhey

Healthy chickenpox scabs can in fact leave inconspicuous scars, but the more you scratch them, the worse the scar will become.

Healing any clean wound starts with a growth of scaffolding tissue to contain the damage. Next the scar tissue begins adjusting its structure to its functions. In small, clean lesions of simple shape, the scar tissue may adapt so neatly that one hardly notices the mark that is left behind. However, scaffolding tissue cannot form so neatly on larger wounds, so they leave more conspicuous scars that may take years to shrink or may require plastic surgery.

Interfering with scar formation, by repeatedly scratching a scab, for instance, aggravates scarring. Also, when the tissues are infected with germs such as the chickenpox virus mentioned in the question, the pathogens not only interfere with tissue growth, they attract the leucocytes that can form pus. Left alone, healthy leucocytes cleanly kill both the pathogens and infected tissue; they cordon off the pustule until everything dries and sloughs off, so that scar tissue can form neatly afterward. But interference with this process, such as scratching, messes up the pustule structure, exposes more tissue to pathogen and leucocyte damage, and thereby creates a larger and more unsightly scar.

Jon Richfield

LONG LIVE VIRUSES

Plenty of viruses will shorten your life or kill you. Are there any that make you live longer? After all, it is in the interest of the virus to keep its host alive and ensure maximum propagation to other hosts. If there are no such viruses, what is the flaw in my suggestion?

Graham Lundegaard

Whatever the benefit to a virus of a long-lived host, there is no guarantee that hosts have mechanisms for increasing longevity that the virus could exploit. Nor need host longevity be an advantage. The interests of the virus depend on its life-cycle strategy, and some viruses are only released on the death of the host. In fact, many parasites actually force their host to attack other potential hosts or to be killed or eaten so that the parasites (viral or otherwise) are passed on.

On the other hand, many of the germs that live in or on the bodies of all animals, including humans—our normal microflora—are not simply parasitic, but contribute important protective, stimulant, or nutritional effects. In some ways this can be seen as a form of life extension because removing them could have drastic, possibly fatal effects. Extreme examples include endosymbiont root fungi in orchids and mitochondria in our cells. Without them orchids and humans would not survive.

Peter Christin

There is a school of thought that claims that all parasites are evolving to a point at which they become innocuous inhabitants of their host, and that the age of a species can be deduced by its virulence (a measure of damage done to the host). If damage is negligible, the species can quite often be considered to be an old one. However, this position is increasingly being viewed as inaccurate.

If endangering the life of the host is a result of a strategy that leads to increased probability of viral transmission, then the strategy may not, as your correspondent suggests, be counterproductive. The acceptable level of virulence of a virus is dependent upon its relationship to transmission. For example, diarrhea can kill, but it is an excellent way of ensuring transmission.

As for viruses that make you live longer, it should be noted that some viruses seem to have become integrated into their host's chromosomes, and are therefore inherited. Presumably they confer a survival advantage upon the host, or at least do no harm.

Paul Dollin

It is sometimes in the interest of viruses to keep their host alive long enough for the virus to make use of its cellular mechanisms to reproduce itself and spread itself through contact between the current host and other potential hosts. For this reason the myxomatosis virus rapidly became less virulent when intentionally released into the Australian rabbit population. Not only were the rabbits evolving an immunity, but natural selection favored viruses that did not kill thc host so rapidly, or even left it alive.

This is an example of a virus increasing the time it takes for its host to die, not increasing the host's longevity over that of an uninfected animal. A virus's main aim is to replicate as many times as possible as quickly as possible. To do this it takes over the host cell's DNA replication and protein synthesis capabilities and eventually breaks the cell apart to release multiple copies of itself. Clearly this cannot be good for the organism. So while immediate death of the host may not be desirable for a virus, the life cycle of viruses is primarily destructive and so will be harmful to the host animal in the long term.

Johanna Rayner

I don't know of any viruses that make humans live longer, but here at Plymouth Marine Laboratory we are working on a giant virus that infects the marine alga *Emiliania huxleyi*. The virus, known as EhV-86, contains seven genes involved in the biosynthesis of ceramide, a compound that plays a role in controlling programmed cell death, or apoptosis. We believe this virus keeps its host healthy for as long as possible, allowing it to make hundreds or thousands of copies of itself. Instead of the viruses accumulating in the host cell and being released when the cell bursts at the end of the infection cycle, the viruses actually pop through the membrane one by one, thus keeping the host cell healthy and intact.

Mike Allen
Plymouth Marine Laboratory, United Kingdom

Some viruses are being used in the fight against cancer. For example, researchers from the Hebrew University of Jerusalem are targeting brain tumors using a variant of the Newcastle disease virus, which usually afflicts birds.

Likewise, genetic engineers use harmless viruses to carry desirable genes into cells. For example, scientists have tried treating familial hypercholesterolemia, a genetic disease, by

infecting liver cells with a virus. The virus inserts a crucial gene that makes the liver cells produce a chemical sponge that controls harmful cholesterol.

Trying to hijack viruses for our own ends goes back at least as far as the close of the eighteenth century, when Edward Jenner discovered that milkmaids infected with relatively mild cowpox were immune to virulent smallpox. This eventually led to widespread vaccination using the closely related vaccinia virus.

Mike Follows

WATER CURE

New Scientist's book Does Anything Eat Wasps? had a question about athlete's foot. I used to suffer this to excess until my father told me to pee on my feet when I shower. I've never had it since. How does this "cure" work?

Tony Male

Athlete's foot (or tinea pedis) is a skin infection caused by several different ringworm fungi. The most common culprits are *Trichophyton, Microsporum,* and *Epidermophyton.* Ringworm fungi grow best at 95°F, so when skin stays moist and warm, ringworm thrive and infect toenails and upper layers of the skin, digesting the keratin there. Ringworm fungi can be picked up from floors or clothing, and can live anywhere on your skin, but without suitable conditions—the warm, moist environment of a sweaty socked foot is ideal—they will not cause a noticeable infection.

Most treatments involve keeping the skin drier or applying

fungicide, and at first glance urine appears to have antifungal effects. While mostly water, it does have organic products such as urea, uric acid, and creatinine, some of which are antifungal or antibacterial. Urea paste is used to treat nail fungal infections, because it denatures proteins in the epidermis.

However, peeing in the shower is probably not providing the cure, because urine only contains around 18 percent urea—the pastes have to be at least 40 percent to work—and must be left on the skin for several days. The urea in the paste softens the skin and allows antimycotic medicine that kills the fungus to penetrate. The urea in your urine is not concentrated enough, and if you are peeing in the shower, it is being washed off too quickly to have any effect.

So the most likely answer is that knowing that you have peed on your feet you will rinse and/or dry them especially carefully (maybe even subconsciously), which removes the dead skin cells the fungi use as food.

Jo Burgess
Department of Biochemistry, Microbiology, and Biotechnology
Rhodes University, South Africa

DRAWING THE STING

For years I have reacted badly to horsefly bites, but recently I have discovered that holding a hot mug of tea against the bite gives instant and prolonged relief: the red lump diminishes and the itching stops for hours. Why would heat have this effect, and would it also work on midge bites or wasp stings?

Kim Russell

Heat remedies have a long history and there are now devices that treat bites and stings via heat. But take care: some venoms can cause serious allergy and injury, and there is always a risk of infection, so if in doubt, get advice from a doctor.—Ed.

If you hold a hot mug of tea against your skin, the skin will turn rosier. This is not the skin cooking, but your blood being diverted to the area to try to correct the temperature anomaly. This increased blood flow will wash the poison of the sting away from its original site and carry it off in a much diluted form. Your body can then deal with the poison in a less vigorous and noticeable way than in the original red lump.

Michael Strawson

When an insect bites, it often injects a substance to prevent blood clotting, while the pain and inflammation are caused by venom. In the case of many insect stings and bites, the anticlotting agent or the venom is thermolabile and this breaks down and becomes innocuous when heated. The irritating material in a horsefly bite may well be thermolabile if a hot mug of tea applied to the bite relieves the lump and itch. The same action would probably also work with midge bites and wasp stings.

Mo Laidlaw

This phenomenon is well recognized and forms the basis of the emergency treatment of weaver fish stings, in which the limb is immersed in water as hot as the victim can bear. There are probably three mechanisms at play. First, just as with cooking egg white, heat denatures the protein in the sting, which therefore becomes ineffective. Second, blood flow to the area is increased, removing the toxin from the tissues more quickly. Finally, the heat utilizes the gate-control theory of pain, in which an alternative sensory signal or stimulus will block pain in a similar way to the relief found when scratching an itch.

Geoff Sharpe

5 PLANTS AND ANIMALS

BEE PLUMP

Can insects get fat?

Walt Malker

If, by getting fat, the questioner means obese, the answer is no. All insects undergo some sort of metamorphosis, passing through larval stages before becoming an adult. The adult, or "imago," stage is relatively short-lived and very often adult insects do not feed at all. Mayflies (of the order *Ephemeroptera*) and many silk moths (of the family *Saturniidae*) are some examples. They neither have the time nor the inclination to feed and get fat.

Those imagos that can feed are constrained by their inflexible exoskeleton. They have no means to expand this to take on excess fat. Incidentally, carcasses of death's-head hawk moths that have been stung to death are often found in beehives, where they have made a vain attempt to feed. Their proboscis is too short to enable them to extract nectar from flowers, but long enough for them to consume honey if the bees in the hive would let them.

The larval stages are similarly constrained. To grow they have to molt their exoskeleton periodically. Internal fluid pressure splits the exoskeleton and the insect expands into a new one it has grown underneath, which remains flexible just long enough to accommodate its increased size. Insects cannot keep doing this indefinitely; each species is limited to a predetermined number of molts. If they find abundant nourishing food, they will go through their molts quickly. If the opposite is the case, or if their food contains limited nourishment, they will take a relatively long time to finish growing. Either way, once the molts are completed, they begin metamorphosis into the adult stage. They do not continue to get fatter and fatter and, in fact, the reverse can be true—those larvae unable to find sufficient food may begin metamorphosis early, skipping one or two molts. This produces a normal, if somewhat smaller than average, imago.

However, insects do store up a great deal of fat at the larval stage. Silk moths are generally large insects. Because they do not feed as adults they must have considerable fat reserves to enable the males to track down a female and mate and, in the case of the females, to produce and lay a large number of eggs while sustaining their metabolism for a week or more. But this is the normal situation; they are not in any way obese.

Terence Hollingworth

Insect life cycles do not lend themselves to the concept of obesity as we know it. Most species accumulate food as fast as they can, but once they have enough, they enter their next stage of life, or reproduce and die. There are exceptions, but few can afford to get too fat—anything that interferes with

their bodily function prevents reproduction—so what they cannot use, they dump. For example, sap-sucking insects get far too much sugar from plant sap, but instead of becoming uselessly fat, most dump the excess as honeydew or convert it into waxy armor. Using hormones to prevent insects from maturing may make them larger and fatter, but prevents their breeding.

Still, healthy insects do accumulate some fat. Their internal "fat bodies" are special organs crucial for storage, hormonal control, metabolism, growth, overwintering, fuel for traveling, yolk production for eggs, and so on. Accordingly, many insects, such as locusts and termites, though not technically obese, are prized as fat-rich foods. As you may have seen on TV, termite queens of most species accumulate huge fat stores to support their role as virtually continuous egg-laying machines.

Jon Richfield

With my colleague James Marden I have described (among other symptoms) infection-associated obesity in a dragonfly species. Infected dragonflies show an inability to metabolize fatty acids in their flight muscles and so build up lipids in their thorax, leading to a 26 percent increase in thoracic fat content. The suite of symptoms caused by this infection includes decreased flight performance and decreased reproductive success in male dragonflies (Rudolf J. Schilder and James H. Marden, "Metabolic Syndrome and Obesity in an Insect," *Proceedings of the National Academy of Sciences,* vol. 103, no. 49 (2006), pp. 18805–9).

Ruud Schilder
School of Biological Sciences
University of Nebraska

PARCHED PERCH

Do fish get thirsty?
Jack Bennett

Well yes, at least some of them do, so long as we leave aside the subjective human feeling of "thirst." There is also a substantial difference between fish in seawater and fresh water, and we need to consider the possibility of the thirsty shark.

Bony fish, known as teleosts, have a salt concentration in their bodies that is not dramatically different from that of land-dwelling vertebrates. This means that the teleosts of the sea—marine fish—live in an environment with a much higher salt concentration than is present in their blood. Their relatives in fresh water are in the opposite position.

Water tends to move along concentration gradients through water-permeable biological membranes like those that shield most organisms from their environment—a process known as osmosis. Therefore, marine fish, which have a low salt concentration compared with that of seawater, will constantly leak water through their body wall—especially through the thin and permeable gill epithelia. To replenish this lost water, marine fish need to drink, so it would be easy to argue that they become thirsty. The surplus salt they ingest by drinking seawater is excreted by specialized cells located in the gills.

Freshwater fish, on the other hand, are unlikely to become thirsty. Because they live in a more dilute environment, they have the opposite problem: water flows inward and dilutes their blood. The freshwater fish therefore need to excrete excess water, which they do in much the same way we do, via a dilute urine.

So we can see that marine fish get thirsty and drink, while freshwater fish avoid drinking but pee a lot.

Finally, sharks, dogfish, rays, and skates—which are cartilaginous rather than bony and are called elasmobranchs—are also marine fish (with a few Central and South American freshwater exceptions). Although the concentration of inorganic salts in their blood is not dramatically different from that of marine teleosts, they have little or no osmotic gradient between blood and seawater. This is because they retain organic molecules instead, the main ones being urea (carbamide) and trimethylamine oxide (TMAO). In this way, the cunning sharks avoid an osmotic water flow from their body surfaces, and may not be very thirsty.

Stefan Nilsson
Professor of Zoophysiology
University of Gothenburg, Sweden

FEARFUL FELINES

Cats all seem to like fish, so why are they unwilling to swim?
Tom Lorkin

I like fish too, but you won't catch me on a trawler, let alone in the water. For one thing, like most cats, I simply cannot swim well enough to catch any meal that swims away from me. While otters, seals, and other aquatic mammals and reptiles swim after the fish they hunt, there are comparatively few animals like them. Some cats fish actively, but they do so from the bank, or leap on fish in shallow water. Whether cats in a particular region catch fish depends on their having learned the skill. Where fishing is good, cats of all sizes from wildcats to jaguars may snatch fish, and the Southeast Asian fishing cat apparently does so frequently.

As for being unwilling to swim, cats vary. Some actually like it. I have seen the hilarious sight of a tabby lowering itself into a swimming pool tail-first to avoid getting a nose full of water, swim around to cool off, then head for the steps. The breed commonly called the Turkish Van is well known for enjoying the occasional summer dip.

Jan Rhode

Cats are quite capable of swimming if they have to but may dislike it because of its effects on their fur. A cat's fur is effective insulation from both the cold and the heat, thanks to the way it lies on the cat. If a cat gets soaked, the fur becomes waterlogged and the cat can lose body heat to the extent that it becomes hypothermic. However, while a cat will seek shelter in the rain, a little damp does no harm because the top layer of fur is water-repellent and rain just bounces off. For this reason, it's not a good idea to dry a moderately wet cat with a towel because water will get through the water-repellent layer to the more absorbent hairs below. If a cat is really wet, it's best to dry it with a hair dryer on a very low setting. However, most cats are frightened of this, so letting it sit in front of a fire is probably better.

Cats can also fish. About twenty years ago my family had a cat that regularly brought back bullhead fish. Cats sit on the riverbank and when a fish comes into range the cat yanks it out with extended claws and throws it over its head and clear of the water. The fish is then helpless and the cat has its meal or trophy.

Charles Stuart

Snow leopards, lynx, and other species from cold environments avoid getting wet because water compromises the ability of their fur coats to keep them warm. On the other hand, lions, tigers, jaguars, and other species that live in hot habitats often take a dip to cool off. It is thought that the Turkish Van, which hails from the region around Lake Van in eastern Turkey, took to swimming to escape the scorching heat. This swimming cat has dispensed with the undercoat that most cats have and its fur has a cashmere-like texture that makes it water-resistant. The fishing cat (*Prionailurus viverrinus*) from Southeast Asia has gone one step further and dives into water to catch fish.

Fishing cats have been reported to attack ducks from under the water.

While a Turkish Van can go for a dip and come out relatively dry, most domestic cats hate getting wet, possibly because they must spend hours putting their fur back in order. However, some domestic cats will happily join their owners in the shower or play with a dripping tap.

Mike Follows

On one occasion my cat swam out to my fishing boat, a distance of about a hundred yards, presumably for company and a feed of sardines. Her swimming style is similar to a doggy-paddle. She only breached the surface when taking a breath (in a similar manner to a seal). On another occasion, we were netting for bait, and she swam behind the net, attacking fish that were caught in it.

I guess some cats swim and some cats don't like fish.

Richard

CLOSING TIME

Why do some flowers close at night? What is the evolutionary advantage of doing this, and why do only some plants bother to do so?

Craig

When flowers close temporarily for the night they are effectively in standby mode, protecting their delicate reproductive parts and pollen while they are not in use. The pollen is isolated from the dew that forms during the night, keeping it dry so that it can be dusted onto a passing insect the following day. Indeed, some flowers remain closed until some time after dawn, and only reopen when the day is warm enough for the dew to have evaporated.

Closing the flowers also helps to protect against nighttime cold and bad weather. As well as closing their petals, some plants also close the tough surrounding structures, called bracts, to protect the flower against plant-eating insects. Keeping the pollen dry while limiting access for plant-eating insects—and the fungi and bacteria that they carry—also means that the pollen is less likely to spoil.

Ultimately all these adaptations minimize wastage of pollen or damage to the flower. This kind of economy is a particular advantage for plants that live in what might be termed stressful environments with limited resources, where they would be hard-pressed to produce new flowers and so must protect their existing investment. Species that have evolved to make the most of lusher habitats in which more resources are available are not so frugal: their strategy is to produce fresh flowers when needed rather than maintaining, and closing, the ones they already have.

Some plants have adapted the movements of their flowers to odd ecological situations. For example, certain bat-pollinated wild pineapples—members of the bromeliad family—do the exact opposite of most flowers by opening their flowers at night and closing them during the day. This is to protect them from weevils, which are most active during daylight hours.

The mechanisms that flowers use to close their petals are essentially the same as the ones they use to open out in the first place, but are not the same in all species. Those of the *Kalanchoe* genus open their flowers by growing new cells on the inner surface of the petals to force them outward, and on the outside of the petals to close them. Gentian flowers use the expansion and contraction of the cells that form the petal. This type of movement is controlled by genes that switch on or off in response to changes in temperature or the amount of light, and are regulated by an internal clock. The genes regulate the amount of sugars in the petal cell sap; larger amounts cause water to enter the petals by osmosis, pressurizing them and opening the flower. In effect, when these flowers close they do so by wilting.

Simon Pierce
Department of Structural and Functional Biology
University of Insubria, Italy

There are a number of reasons for the differing approaches of various flower types. Long-blooming flowers close and re-open repeatedly, whereas other flowers may drop their petals after a single day.

Long-lasting flower heads of many members of the *Asteraceae* family, for example, and flowers of some members of the genus *Mesembryanthemum* (their very name means "midday flowering") protect their gonads from nighttime dews or frosts but open in sunlight. Where sun, dew, frost, wind, or insects are likely to damage exposed reproductive organs, closing may be advantageous during times when flowers are unlikely to attract pollinators. Analogously, many moth-pollinated flowers release their fragrance only at night, so avoiding waste during daylight.

Some flowers open and close so punctually that a once-popular gardening fashion was to plant flowers in sectors of a bed resembling a clock face. These were planned so that flowers opening in each sector matched the position of a notional hour hand on the clock. In season, all being well, one might actually be able to tell the time by consulting one's flowering clock.

Of course, such nonoverlapping timed opening reduces competition for pollinators, and bees soon learn to concentrate on the most rewarding plants at the times when they open reliably. Arbitrarily visiting assorted flowers would be less efficient for both the bee and the plant, increasing the wastage of pollen on unrelated plants.

Jon Richfield

BOVINE CHALLENGE

How long would it take an average cow to fill the Grand Canyon with milk?

Nicola Stanley

There are many answers to this question, ranging from the pedantic arguments over the definition of an average cow to the defeatists' pronouncements that the stench of sour milk

would be too great. In the end we decided to treat the answer in its purest form and calculate it based on the volume of the canyon and the average milk output of dairy cattle. A surprising number of our correspondents calculated an answer similar to the first one below.—Ed.

Obviously the first job would be to divert the Colorado River, which would otherwise interfere with the process. Second, the canyon would need to be dammed to retain the milk. Third, because this is a desert environment, huge refrigeration capacity will be required to prevent the milk from turning to cheese. And finally, to prevent loss of liquid by evaporation, the canyon will need to be hermetically sealed.

So, preparation complete, let's wheel in Daisy, the average cow. In the United Kingdom, average milk yield per day per cow is in the range of fifteen to twenty liters (four to five gallons). So let's settle on 17.5 liters. The canyon is 446 kilometers (277 miles) long by an average of 16 kilometers (10 miles) wide and 1.6 kilometers (1 mile) deep, which gives a volume of about 10 million billion (10^{16}) liters (3×10^{15} gallons). So by simple division Daisy would take about 1.8 million million (1.8×10^{12}) years to fill the canyon. This assumes the canyon has a rectangular cross section; for a triangular cross section, the time would be halved.

Now, suppose you don't want to wait three hundred times the age of the planet for your canyon full of milk. Instead, you could divert the world's entire milk production to the canyon. This adds another requirement—a milk-pumping infrastructure of epic proportions—unless you choose to use dried milk, which would be cheaper to transport, and then rehydrate it with water from the river. The UN Food and Agriculture Organization estimates that global milk production in 2004 was 504 million tons, which is equivalent to 489 billion liters (129 billion gallons), giving an estimated fill time of only about twenty thousand years—still a pretty long job.

Jon White

It all depends upon the size of the tanker truck the cow chooses to drive, the time it would take to drive from the milk distribution point, the inflow and outflow of the tanker truck, the ability to change the absorption and evaporation rates of the milk, and the ability of said cow to effectively block the exit route of the Colorado River.

Other considerations, of course, would be whether the cow works an eight-hour day or 24/7, and whether she ever has a day off. In a tangential vein, what subsidies would the U.S. government be giving to the dairy farmers? This could be the making of another watershed in reality TV.

Bob Friedhoffer

NO DEAD PARROTS

There are billions of birds worldwide, so why is it that you rarely, if ever, see a dead one?

Maurice Boland

The fate of a bird carcass, or indeed that of any other animal, depends largely on the cause of death. If the bird was killed by a predator, it will probably be eaten immediately and there will be nothing left to see except for the odd feather.

Animals that are sick generally hide somewhere quiet and isolated. So the bodies of birds that die from disease or from old age will be in hard-to-get-to places and will most probably be eaten by ants and other insect scavengers before you chance across them.

The only case remaining, and the main reason we humans see dead animals, is when the animal is killed accidentally by something with no interest in eating it. The most obvious example of this is roadkill, although most birds are fairly fast-moving and agile, and so are less likely to be hit by cars than slower-moving ground animals. However, on a recent trip to central Queensland my parents saw several dead wedge-tailed eagles by the side of the road. These majestic birds, which can have a wingspan of 2.5 yards, are unfortunately rather cumbersome and slow to take off. Therefore, when they land on roads, they are often killed by vehicles.

Ironically the eagles had been attracted to the roadside by the carrion of other roadkill.

Simon Iveson

For each dead bird there are, fortunately, plenty of burying beetles of the family *Silphidae* that will promptly fly to the scene from a considerable distance away, attracted by the enticing smells. So tough is competition for these tasty morsels that sometimes beetles carry mites on their bodies that upon arrival promptly alight on the corpse and start ridding it of the eggs of blowflies or other faster-eating scavengers (see http://bugguide.net/node/view/5957/bgimage).

This buys the beetles some time while they scrape away under the bird, which will soon sink into the ground and disappear.

Once they have located a corpse, a couple of beetles will soon mate and start preparing the nest for their offspring. With the help of mouth and anal secretions they make a "brood mass" with the bird's flesh and tend it so that by the time their eggs hatch it will still be in pristine condition for their larvae.

During the early stages of the larvae's life the parents will feed them with regurgitated bird flesh much as birds do for their nestlings. This form of parental care is very rare among such nonsocial insects, but burying beetles will look after their larvae until they are ready to pupate in the soil. By that stage there will be not much left of the bird at all.

Maria Fremlin

Dead birds accumulate in some areas, such as the edges of lakes, estuaries, and shorelines—this is why avian flu surveyors focus on places like this—and also along roads, where millions of birds die.

Remains are mostly quickly scavenged, but in areas where game birds are released in large numbers, roads are often covered in dozens of casualties. Sometimes these can be useful for stocking one's fridge, but one must question the ecological impact (and perhaps ethics) of large-scale autumn releases of more than twenty million pheasants and red-legged partridges in the United Kingdom.

I recommend a visit to the estate roads of Scotland for an instructive "dead bird" experience.

Ian Francis

BARKING UP THE WRONG TREE?

My four-year-old daughter asked me if her dog knows that it is a dog. Does her pet realize it is different from us or does it think that we're just odd-shaped dogs or, indeed, that it is a particularly impressive human being?

Celia Denton

The question calls to mind cyberneticist Stafford Beer, writing in a 1970s edition of *New Scientist*: "Man: 'Hello, my boy. And what is your dog's name?' Boy: 'I don't know. But we call him Rover.' "

The boy's reply reveals his belief that his dog has a mental image of itself (which he assumes to include a name), but at the same time confesses his inability to penetrate the dog's psyche. And he's right: the short but unsatisfying answer to the question above is that we don't know what goes on in a dog's mind.

We can, though, make a reasonable stab at it. For a start, it's clear that when individuals of different species interact, judgments of sameness or difference are simply not part of the story. That the tiny reed warbler heroically feeds the gigantic cuckoo nestling, despite the obvious (to humans) fact that it cannot possibly be a warbler, indicates that the bird isn't operating according to any concept of warbler-ness or cuckoo-ness, but purely to one of this-thing-needs-to-be-fed-ness.

Dogs have specific responses to things-that-dogs-can-eat (such as rabbits) and things-that-can-eat-dogs (such as lions), and also to potential mates or rivals, and to offspring. Other than that, they resemble humans in viewing a variety of creatures of whatever sort as potential social companions or friends. Indeed, that is why you have a dog—and the dog tolerates you—in the first place. As with humans, the establishment and maintenance of social bonds is key to dogs' way of life. I think your dog sees the questioner as her friend in much the same way that the questioner sees the dog as hers. Lucky dog, lucky you.

Angus Martin

Canids in general start to develop social relationships when their eyes and ears open at about two weeks of age. During the critical period between two and sixteen weeks, puppies learn the social rules that will shape their behavior for the rest of their lives, including recognition of conspecifics and appropriate mates. The famous ethologist Konrad Lorenz, when studying greylag geese, found that sexually mature geese raised by a human "mother" tended to direct their courtship behavior toward humans rather than toward other geese.

In dogs, this same confusion can be seen in the way dogs direct social dominance and play behaviors toward humans—in effect, treating people as if they were dogs. Likewise, livestock-guarding dogs, such as those protecting sheep, are trained for their jobs by removing them from their mothers at just a few weeks of age and allowing them to grow up with sheep as their companions. The sheep are then forever recognized as family and are socialized with and protected as such.

After sixteen weeks, this period of rapid learning and adaptation ends, and the social skills the dog has are pretty much set for life. This is why it is so important for puppies to have intimate contact with people from the time they are born. Traditionally, we adopt pet dogs when they are eight

or nine weeks old, right in the middle of this period of social development, and proceed to lavish them with attention and experiences through to the end of that sixteen-week period. The result is that the dog in the question above sees nothing at all odd about her tall, hairless packmates.

Julia Ecklar

NIGHT LIGHT FLIGHT

If moths are nocturnal, why are they so keen to fly toward light?
BBC Radio 5 Live listener

Moths are not exclusively nocturnal. Many species are active by day as well, and many are exclusively diurnal.

Insect behavior is almost entirely instinctive, as the famous nineteenth- and twentieth-century entomologist Jean Henri Fabre showed us. He noted that a digger wasp of the genus *Sphex* would sting a caterpillar and carry it, paralyzed, to the mouth of the burrow she had just excavated and deposit it at the entrance. She would then enter the burrow, presumably to check that no unwanted occupant had taken up residence there in her absence, emerge again, collect the caterpillar, and take it inside to lay her egg and close the burrow. After observing this, Fabre began to move the caterpillar a little distance from the hole once the wasp was inside. The wasp would collect the caterpillar and then repeat the inspection process.

Fabre was never able to break this routine. The wasp had evolved to behave this way and it was impossible for her to reason out anything different for this unusual set of circumstances. In the same vein, moths have undergone millions of years of evolution without ever having to deal with artificial lights at night. The few thousand years or so during which we have presented them with the problem is too short for them to have evolved appropriate behavior.

Richard Dawkins, in his recent book *The God Delusion* (New York: Houghton Mifflin, 2006), presents us with the problem of moths drawn to a candle flame. His solution is the old glib explanation that the moths are trying to use the flame to navigate, mistaking it for the moon. The idea is that a moth sets its course according to the position of the light, so it will have to keep turning toward it to maintain the same relative heading, and the path it will take will be a spiral leading inevitably into the flame. This explanation does not tell us, however, why it is that in many species only males are thus attracted, and in a few, only females.

What is more, if moths need to navigate, they must be from a migrating species. Yet most of the time such moths are not migrating. Indeed most species do not migrate at all and thus have no need of navigation. Moreover, all groups of insects display the same behavior: flies, wasps, hornets, mayflies, and caddises are all drawn inexorably toward flames, although many of these insects are normally diurnal and the majority rarely, or never, migrate.

So are they navigating to find food or a mate? At night in summer, male moths use scent to decide their heading, not light. Total cloud cover makes no difference to their behavior. They move into and across the wind, hunting for telltale pheromones which will lead them to a female or for the scent of flowers to enable them to feed. Females, for their part, stay still until after mating and then go looking for the scent of

plants that their larvae can feed on in order to lay their eggs on them. They don't need the moon, stars, or candles to do this.

I have spent thousands of hours sitting by light traps observing insect behavior and I feel, for the most part, it is pure accident that they stumble upon the light. Many can be seen to fly straight past without deviating one iota. Others fly into the lighted area, land, and stay still, as they would if it were daytime. Different species seem to have different sensitivities, the most sensitive ones alighting the farthest from the light. Still other moths circle the light, never bumping into it.

Those that do fly toward the light often do so in a wild and confused manner. They seem disoriented and confused by a bright light rather than attracted to it. Just like Fabre's interference with the caterpillar, the light seems to trigger inappropriate behavior, because the insects have no mechanism to deal with it.

There is only one type of observed behavior that seems appropriate, and that is moths moving toward a lighted window at night. Any insects trapped in a room will fly to the window. Instinct tells them that in order to escape from a confined space they have to fly to where it is lightest, and even at night it is lighter outside than inside a cave or a cavity where the moth has been hidden during the day. The moth finds its way out by flying toward that light. So a moth would not be able to distinguish between a bright light source and an open space.

It could well be this mechanism that is operating when moths fly to bright lights; however, the many other factors listed above seem to counter this view.

Terence Hollingworth

I have seen this question asked many times and answered in as many ways, but I have yet to find an answer as attractive as that which I first read in Ian McEwan's novel *Atonement* (New York: Nan A. Talese Books/Scribner, 2002), now a major movie, which is set in the 1930s and 1940s. Incidentally, despite its more historical setting in the novel, I believe the theory originates from 1972 and Henry Hsiao, a biomedical engineer. Put simply, nocturnal moths fly toward dark places and with only simple light-sensing apparatus they perceive the area behind and beyond any point of light, such as a lightbulb, as being the darkest around. Sadly, this idea does not seem to be shared by entomologists and has never been confirmed experimentally.

Rob Jordan

SWEAT ON THE WING

Do pigeons sweat? If not, why not?

Class 3L, Hungerford Primary School, London

Only mammals have sweat glands—so no, pigeons do not sweat. Nor do mammals such as cats, whales, and rodents, which have lost most or all of their eccrine sweat glands—the ones that we use in shedding heat—while birds never developed them. In sweat-free mammals the kidneys deal with the excretory functions of sweat, and flushing or panting is how they cool themselves down. As another example of evaporative cooling, an over-hot cat not only pants but also moistens its fur with saliva.

Bird skins are dry. However, as birds have body

temperatures that are generally several degrees higher than those of mammals, they do not need the same capacity to lose heat.

When they do need to lose heat, they can raise their down feathers to cool the skin by ventilation. To conserve heat, they flatten them. Beyond this, panting through open beaks causes evaporative cooling: hence the Afrikaans expression: "So hot the crows are yawning" ("*So warm dat die kraaie gaap*").

Finally, on very hot days, many kinds of birds, including pigeons, enjoy a bath.

Byron Wilson

Storks, cormorants, and vultures indulge in urohydrosis: they literally wet and thus cool themselves by urinating down their legs. Because birds do not urinate and defecate separately, everything comes out together, which makes bird droppings very watery. The heat required to evaporate this liquid from the surfaces of the legs cools the blood, carried close to the surface of the legs by a network of veins.

Before condemning these birds for their unappealing party trick, it is worth adding that bird droppings contain uric acid, making it an effective antiseptic—very useful for vultures that spend a lot of their time trampling over rotting carcasses.

Mike Follows

WHEELS OF FORTUNE

Could hamster power be an environmentally friendly answer to the impending energy crisis? How many hamsters running on wheels would it take to provide energy for a house or a factory?

Catherine Hetherington

Thanks to everyone who pooled their physical, chemical, and mathematical skills to calculate the impact of national or global hamster power. Sadly, no two people came up with the same figures. There is substantial disagreement even over the average weight of a hamster. Nevertheless, the conclusion is the same: we are all hamster-power skeptics.—Ed.

Let's assume a hamster weighing 1.7 ounces (48.2 grams) can run up a thirty-degree slope at two yards per second. This corresponds to a power output of half a watt. If it delivers the same power when running in a hamster wheel, we would need 120 hamsters working flat out to light a sixty-watt bulb.

The average hamster probably doesn't spend more than 5 percent of its life running in its wheel, so already we need a brigade of twenty-four hundred hamsters just to light our bulb. It gets worse. The average UK household consumes in excess of eighty gigajoules of energy per year. This is equivalent to a constant power consumption of about 2.5 kilowatts. Each house would need a hundred thousand hamsters. Multiply this by the number of households in the United Kingdom and we would have an environmental and economic disaster.

In addition, we would need to employ an army of animal behaviorists to devise Pavlovian tricks to get the hamsters onto their wheels in response to surges in demand. And given that hamsters are nocturnal, this would force politicians and lawyers to debate animal welfare. The United Kingdom alone

would need to employ everyone else in Europe to feed and care for its hamster population.

Perhaps we should let humans run on treadmills. It would not produce much electricity but we might end up with less of an obesity problem.

Mike Follows

Hamsters running on wheels cannot relieve the impending energy crisis because animals are not energy sources—they are energy consumers. It would be more efficient to simply burn their food in a furnace and use the power output from that (a fact overlooked in *The Matrix* films, where humans are used as thermal energy sources by sentient computers).

John Woods

According to the CIA website, the estimated global electrical energy consumption in 2003 was 15.45 trillion kilowatt-hours. To produce that kind of energy in ideal conditions would require around 1,458 billion hamsters. Hamsters have an average life span of 2.5 years, meaning that if we had switched to hamster power in 2003, we would already have more than two billion tons of depleted hamster, and many backyard funerals. The environmental and socioeconomic impact of this would be devastating. So it is my duty as a pseudotechnician to decree that this is another energy source best left to fiction.

Ben Padman

The question is not whether hamster power is an environmentally friendly energy source, but whether it would be welfare friendly. As a veterinary student, I have spent some time looking at research into "the running wheel phenomenon."

It is clear that captive hamsters are highly motivated to use running wheels. What mystifies researchers is why.

There is controversy over whether running-wheel activity is a stereotypic behavior—repetitive, invariant behavior with no obvious function (likened by some to obsessive-compulsive disorder in humans) which results from a sub-optimal environment. Even if running-wheel use turns out not to be a stereotypy, there is further debate as to whether it corresponds to poor welfare, because such behaviors may merely be a way of coping with captivity.

However, what has been found is that when hamsters are given bedding that is thirty inches or more deep, which lets them indulge in natural burrowing behavior, their use of running wheels drops dramatically and the performance of other stereotypies such as wire-gnawing ceases altogether. This suggests that we should reconsider how pet hamsters are kept. Perhaps finding a way to harness the burrowing activities of hamsters would be a better solution to the energy crisis.

Sarah Briars

Total world annual energy consumption is about five hundred exajoules (an exajoule is 10^{18} joules). A hamster requires fifteen grams (half an ounce) of food per day. Let's assume that the hamsters eat wheat, with an energy content of fourteen hundred kilojoules per one hundred grams (3.5 ounces). If we assume that they convert the chemical energy of their food into useful energy with the same efficiency that power stations and wood-burning stoves do, some 6.5 trillion hamsters on wheels would be needed to supply the world's energy requirements. On a more manageable mental scale, energy use in a typical house in the United Kingdom is about eighty gigajoules a year, which is the

amount a thousand continuously running hamsters would produce.

The drawback to maintaining 6.5 trillion hamsters is that worldwide they would need thirty-six billion tons of wheat per year, nearly sixty times the world's present wheat production.

This exercise illustrates that the world's energy crisis is not simply due to excessive use of fossil fuels, to be solved by conversion to renewable energy sources. The scale of such a conversion is far too great, and large-scale renewables have their downsides too. Cutting energy use through conservation and, by implication, a change in the way we live, is the main answer to the energy and climate change crisis. This is why politicians find it nearly impossible to confront the issue.

Philip Ward

According to the 2001 Residential Energy Consumption Survey conducted by the U.S. Department of Energy's Energy Information Administration, the average American household consumes about 97 gigajoules of energy per year—about 120,000 hamsters.

And of course, once your energy-producing hamster has expired you'll need to dispose of it in useful and reverential fashion. The companion to this book, How to Fossilize Your Hamster, *explains exactly what to do.—Ed.*

THIRSTY SPIDERS

Do spiders drink water? If not, how do they quench their thirst?

Sarah Cassidy

Yes, spiders do drink water. In the wild, most will drink from any available source such as droplets on vegetation or the ground and from early morning or evening dew that has condensed on their webs. For those kept in captivity, it is a good idea to provide a fresh water source such as a small bottle cap or damp sponge for smaller species, or a small dish for larger species such as tarantulas.

Incidentally, spiders' need to quench their thirst seems to have given rise to the myth that they live in drains. When a spider is in a building, an excellent source of water is droplets left from taps and showers around the plugholes and sink edges. Needless to say, spiders remain trapped in the sink or bath because the sides are too slippery or steep for them to climb, hence their tendency to head for the drain as a means of escape.

Sean Lenahan
Cartmel College
Lancaster University, United Kingdom

Like all other animals, spiders require a regular intake of water. Different species use different methods to quench their thirst. For example, the whistling spider, found in the desert, covers its one-yard-long burrow with a thin layer of silk to keep it humid. Dew or the occasional raindrop is captured using a low, silk-covered mound near the entrance. Many other species, such as the wolf spider, opt for a much simpler strategy by drinking dewdrops in the morning. Some spiders even ingest nectar.

Paul Peng

Many spiders, such as the common garden spider, will devour their web first thing in the morning. In doing this, they consume the water that has condensed as dew droplets on the web. Other spiders, such as the whip spider, can use their pincers to take water into their mouths.

The black widow or the red back do not drink water at all. They get all the fluid they need from the juice sucked out of their prey. Tarantulas, on the other hand, like to drink water droplets that have collected on nearby leaves and foliage.

There are some creatures, including mammals, that do not drink. The name koala is derived from the aboriginal word "no drink." Koalas get the fluids they need from eating the leaves of plants such as the smooth-barked eucalyptus.

Louise Lench

A few winters ago I watched a spider just outside my kitchen window as a small snowflake landed on its web. Normally a spider does not react to the presence of something in its web unless it struggles, so I was surprised to see the spider run to the snowflake. By the time it arrived, the snowflake had melted into a droplet of water and the spider gave every appearance of drinking it. The spider's head was against the droplet, and the droplet dwindled away to nothing.

Norman Paterson

The Australian naturalist Densey Cline once reported a remarkable case of a spider drinking water.

She awoke to find the shriveled body of a dead huntsman spider lying on her bedside table, and in her glass of water was an astonishingly long parasitic worm. She speculated that the mature parasite required water to complete its life cycle and had driven the infected spider to the nearest source of water by inducing a terrible thirst.

Kate Chmiel

COCOON MYSTERY

When an insect is changing inside its cocoon, and has turned to slush, is it alive? And if so, in what way is it alive?
Madeleine Cooke (age seven)

In many metamorphosing insects, the majority of the cells in the body of the pupa do break down and turn to mush, but there are clusters of cells that remain intact. These cells feed on the mush, divide, and go on to develop the legs, eyes, wings, antennae, and so on that we see in adult insects.

It is almost as though the mush is the yolk and the cluster of cells is the embryo of a new egg. In some rare cases, such as fungus gnats, this new embryo can split to form multiple "twin" adults from a single larva. This is called polyembryony.

Martin Harris

An insect undergoing metamorphosis is alive regardless of what state its body may be in. For one thing, the individual cells are alive and are growing and dividing in a coordinated manner to form the organs of the new adult insect. An insect, or any other organism, could not be dead at one stage of its development and alive at a following stage, because the death of an organism is always irreversible.

However, the death of a multicellular organism such as an insect must be defined separately at different levels of organization: the intact body; the organs and tissues; and finally the individual cells. The body cannot survive without organs and cells, but the latter two groups can survive without a body. If you squash a cocoon the larva inside will be killed, but many of its cells will remain alive, at least for a while. Therefore, a multicellular organism can be killed by

destroying its highest level of organization while leaving most of its organs and cells alive. If that were not the case, there would not be such possibilities as human organ transplantations or cell cultures.

Aydin Orstan

What is meant by alive? Is the ball of cells that make up the human blastocyst alive? It cannot breathe, think, or feel pain. But it is alive, like the mush of insect cells in a cocoon. The cells that make up both structures are metabolizing, dividing, and responding to their environment—all hallmarks of life.

Roger Morton

KILLER BUSH

While taking a break on a country bike ride in Australia, I saw an unfortunate insect impaled on the thorn of a low bush. We'd had strong winds in the days beforehand and I can only assume that the insect was blown onto the thorn, which had penetrated the open wing casing before impaling the body. What are the chances of an event like this occurring?

Paul Worden

There are examples of impalement of insect species, most often scarab beetles, on spines or other sharp parts of plants and on barbed wire in the zoological literature in Australia, but only rarely.

It is very unlikely that the wind alone could have impaled the elytron—the modified, hardened forewing of beetles which covers their softer flying wings. That would be like an insect collector attempting to pin a beetle by throwing it at the pin. The elytron is part of the hard exoskeleton and would almost always deflect the glancing blow of a pin or spine. The chances of impalement through an elytron while in flight would appear to be very remote because the flying beetle holds its elytra open, at a wide angle to the body, and they would hinge back toward the body if touched on the outer side by a spine.

A more likely scenario is that the strong winds blew down a twig or branch to which the beetle was clinging and that the extra momentum of beetle plus branch impaled it. Similarly, strong winds might have caused one branch to thrash against another on which the beetle was clinging.

Ian Faithfull
Extension Support Officer
Catchment and Agriculture Services, Victoria, Australia

Dung beetles often impale themselves on the barbs of wire fences while in full flight. In New South Wales I quite frequently see barbs on which insects are impaled.

Toshi Knell

Actually, this event is quite likely in certain areas. The poor beetle probably did not get there by chance, but rather because it was put there.

In Victoria, the culprit is likely to have been a gray butcher-bird (*Cracticus torquatus*). These predatory birds, which are about the size of a small dove, eat large insects, small mammals, and other birds. They skewer their prey on thorns to hold it while they eat. Sometimes they will impale the prey and leave it for a snack later. Suitable bushes near nesting sites can be festooned with victims, including poultry chicks.

There is a record of a butcher-bird returning to its nest to find its three chicks dead after a spell of cold rain. The bird took them from the nest and hung them nearby, returning to eat them a few days later.

Shrikes, which are common throughout Eurasia, Africa, and North America (and also sometimes known as "butcher-birds," although they are not related), have similar habits.

Rob Robinson
British Trust for Ornithology, Norfolk, United Kingdom

AIR TRAFFIC CONTROL

Why don't insects—particularly flies, wasps, and bees—fly in straight lines like most birds do? Their flight patterns seem chaotic and often circular in motion. Surely it is inefficient for them to fly in such a random way?

Mike McCullough

Birds seldom fly in straight lines, in fact, and any self-respecting ornithologist can identify many species by flight pattern alone. That said, birds tend to fly fairly directly to a target in sight, and more complex journeys tend to follow efficient paths based on learned topography.

Insects' flight is much more varied. Their flight patterns partly compensate for their typically poor eyesight, by allowing them to gather more visuospatial information.

When insects do have a path in mind, many fly directly enough, as anyone who has been stung can attest. In threat display, a bee buzzes wildly, but a true attack is like a projectile from a peashooter. In searching, scent-following, mating assembly, territorial patrol, and so on, insect flight patterns are not straight, but they do resemble corresponding vertebrate behavior.

Differences between, say, insects' apparently chaotic swarms and skeins of large birds do not reflect their psychology but rather their size difference, which affects wing motions and the benefits of flying in fixed formation. Flocks of small birds such as starlings or queleas can look very locust-like.

Jon Richfield

Judging insect flight depends on the scale you choose. If you watch a meter of flight and you see the insect traveling in a straight line, you might conclude it does this all the time. But if you observe a kilometer of flying, the picture is quite different.

When insects are searching, their flight is an intermittent combination of long- and short-distance steps where each step differs from the previous one by a small angle. Small steps are the most frequent, while very long steps are rare. This is an optimal searching strategy known as Lévy flight. It seems chaotic but it is not; the angles and the distance follow well-defined statistical distributions. In the case of bees, they fly in amazingly straight lines for kilometers once they know where a food source is. When they aren't flying "randomly," a straight line is the best way to go.

Octavio Miramontes

Insects do not have vision as sharp as that of mammals or birds. The insect compound eye is more attuned to movement and so it cannot precisely position distant objects. As a result, insects tend to take a rather wobbly flight path to navigate to a particular object.

Additionally, many insects navigate using scent. Take the case of a parasitic wasp that is seeking a caterpillar in which to deposit its eggs. In order to locate the caterpillar, the wasp needs to balance the odor signals received by its two antennae. This necessitates a rather wobbly flight path to "lock on" to the source of the scent. When an insect is close to the object it is seeking, this wobble starts to reduce and eventually the insect becomes capable of very precise, rapid actions, such as those of a dragonfly catching a prey insect when both are in midflight.

The "random" movement is a simple product of insect sensory systems. Humans rely on vision to navigate the world—just try locating an open bottle of perfume in a room with your eyes closed.

Peter Scott
School of Life Sciences
University of Sussex, United Kingdom

If readers are interested in how humans can be taught to follow scent trails in the way that dogs and other animals can, then the following story will enlighten them: "Unleash Your Inner Bloodhound—Start Sniffing" (http://www.newscientist.com/article/dn10810).—Ed.

I suspect evolution wasted no time in eliminating those insects which fell easy prey to bats, birds, frogs and the like because of their use of efficient and direct—but predictable—flight patterns. Presumably those insects with variable patterns are those that survived until today.

Peter Tredgett

JUMBOCHOO

Do elephants sneeze?

Robin Rhind

I frequently camp in the bush close to where I live in northern Botswana. By far the most pleasant way to experience the African night is to sleep under a mosquito net rather than in a tent, though you may end up, as I have, being investigated by lions, hyenas, hippos, and elephants, which can be quite exciting.

A friend told me of a time when, sleeping under a net, he woke in the middle of the night and, not being able to see the stars, believed that it had clouded over and might rain. But as his eyes focused more clearly he realized that he was looking up at the underside of an elephant. Being inquisitive, the elephant was sniffing him through the net. Then, suddenly, there was an eruption from the elephant's trunk and my friend's face was covered in elephant mucus! The animal then carefully stepped over the net and went on its way.

Sneezing is an involuntary response that serves to remove foreign or excess material from the nasal passages. Elephants are just as liable to experience foreign matter in their nasal passages as other mammals and presumably sneeze for the same reason as do dogs, cats, and humans.

So yes, elephants do sneeze.

John Walters

I have experienced this at the zoo, feeding the elephants. Some children were also feeding one of them and, instead of holding out a handful of food, one of them held out a handful of pepper.

The elephant reached out, vacuumed the pepper right up its trunk, and what happened next was not pretty. The elephant snorted and wheezed a few times, and then flew into a sneezing "rage" that resembled hurricanes being forced through a flailing fire hose.

The kids were thrown out of the zoo and, thankfully, the elephant was fine once its epic sneezing attack had concluded.

By e-mail, no name supplied

Elephants do indeed sneeze. The thick lining of their trunks makes it more difficult for a substance to irritate the membranes inside, but some African farmers have discovered that chilies are an ideal elephant deterrent. In an effort to protect farmers' crops from elephants, which would otherwise damage or eat their produce, chili seeds are planted around crop fields and dung cakes laced with chili are burned at night.

A charity called the Elephant Pepper Development Trust actually helps develop the use of chili and chili oil in order to induce sneezing in elephants.

Georgia

LUNATIC CACTUS

My Cereus forbesii *cactus flowered last night, coinciding with a full moon. The* Selenicereus grandiflorus *cacti that I had in Bangladesh nearly always flowered at or within a couple of days of a full moon or, more occasionally, a new moon. How is flowering in such plants triggered by the lunar cycle?*

Hugh Brammer

Plants like *Selenicereus* flower at night, when temperatures are low and the creatures that pollinate them are about. A white flower opening at night is highly visible, particularly with a full moon to illuminate it, so nocturnal flowering makes sense in evolutionary terms.

There is also good evidence to suggest that plants sense the length of the nighttime and that these periods trigger flowering. Therefore "interrupting" the night with a bright light such as a full moon could have an effect on flowering in plants. But I know of no scientific studies that have shown this to be the case.

An Internet search provides very few reports of *Cereus* or *Selenicereus* flowering in response to a full moon. So the questioner's observation is likely to be the result of chance.

There are twenty-eight days in a lunar month, and on three of those days the moon will be at its brightest (approximately a full moon). So the plant has roughly a one-in-nine chance each month of flowering at the time of the full moon. I suspect that if it flowers at another time, the observation is not considered remarkable and so does not get reported.

Peter Scott
School of Life Sciences
University of Sussex, United Kingdom

I have a specimen of *Selenicereus grandiflorus* in my greenhouse which produced seven flowers in the summer of 1998. The first flower opened in the night of June 20 or 21 and the others at regular intervals during the following two weeks. The last one opened on the night of July 5. There was a new moon on June 24 and the first quarter was on July 1.

So it is difficult to conclude that the flowering of my cactus was triggered by either the full moon or a new moon.

However, the question aroused my interest and I did some research. I discovered the work of Yosef Mizrahi at the Ben-Gurion University of the Negev in Israel. He and his team have been exploring the possibilities of growing the vine cacti *Hylocereus* and *Selenicereus* as fruit crops: his team have more than 240 genotypes of vine cacti from these two genera in their gene bank.

I e-mailed the above question to him. His response was that while these species flower at different times of the year, he and his team have not observed that the full moon triggers the opening of their flowers, although he admits that his researchers have not been actively looking for such a phenomenon.

Trevor Lea

> There is no greater folly than to be very inquisitive and laborious to found out the causes of such a phenomenon as never had an existence, and therefore men ought to be cautious and to be fully assured of the truth of the effect before they venture to explicate the cause.
>
> —*The Displaying of Supposed Witchcraft* (1677), by John Webster

I have been growing cacti for more than sixty years and have heard this claim several times. Although many flower at night, I have observed no examples of any cactus coordinating its flowering with phases of the moon. I have hundreds of photographs of cacti in flower, some of which have been published in journals. They were taken with a digital camera and are automatically labeled with the date. Having just checked these flowering dates against the phases of the moon, I can tell you that there is no correlation, hence the quote I provide above.

The water loss from large flowers is extreme and so cacti generally keep theirs open for very limited periods. *Micranthocereus purpureus,* for example, opens its flowers after sunset and closes them permanently the next morning, before sunrise.

Jim Ring

SICK AS A HORSE

On a long motorway journey while driving behind a horse-drawn carriage, I wondered, do horses get travel sick? In fact, do we know whether any animals besides humans suffer from motion sickness?

Neil Bowley

Horses are unable to vomit, except in extreme circumstances, because of a tight muscle valve around the esophagus. So it is difficult to know whether or not they feel sick. Other monogastric animals can vomit. Younger cats and dogs frequently vomit during their first car journeys but rapidly become accustomed to travel and no longer suffer sickness. In the United Kingdom a neurokinin-1 receptor antagonist has recently been licensed as a treatment for motion sickness in dogs as it reduces the urge to vomit.

James Hunt

Motion sickness is common among animals, affecting domestic animals of all kinds. A carsick dog is not only pathetic, but messy. In his unforgettable book, *A Sailor's Life* (New

York: HarperCollins, 1956), Jan De Hartog wrote: "My worst memories of life at sea have to do with cattle. Two things no sailor will ever forget after such an experience are the pity and the smell . . . cattle get seasick, and the rolling of the ship terrifies the wits out of them. A seasick monkey or pup may be amusing and easy to deal with, but five hundred head of cattle in the throes of seasickness are a nightmare." He also mentioned horses explicitly. Even fish transported in unsuitable conditions may show signs of disorientation.

Motion sickness is ubiquitous because all vertebrates have organs of balance and they correlate balance with feedback from other senses to stay upright. When movement causes, say, visual information to conflict with balance, the brain of a sensitive individual interprets the disorientation as a symptom of poisoning and a typical reaction is to vomit to clear the gut.

Robert Amundsson

Both Robert Falcon Scott and Ernest Shackleton took ponies with them to Antarctica. On the way they experienced some appalling weather, and both noted how badly affected their animals were. They did, however, perk up when the storms abated. Similarly, Scott's dogs spent most of the storms curled up or howling, suggesting they, too, were suffering. Animals with a similar auditory system to ours would suffer from motion sickness, because it is caused by the confusion of auditory and visual inputs.

Tim Brignall

WOUND LICKING

I know that some animals treat simple injuries by licking them. Are there any animals that, like humans, treat each other's injuries, and do any animals have more sophisticated forms of "medical treatment"?

David Taub

Licking each other's and their own wounds is the most common form of wound treatment for mammals. It is believed that such behavior dates from the earliest days of mammals.

Saliva generally is germicidal and benefits wound tissue, causing little harm to live tissue while helping to slough off or recycle dead tissue.

The habit no doubt developed out of a defensive response to the pain, plus an eating response to bodily fluids and detritus. In fact, when mothers of many species lick sick cubs, if there is no improvement, it can lead to them eating their babies. Distressingly, such disruption may also lead the mother to eat the rest of the litter, especially if they are very young.

Formal hygiene and treatment of illness and injury, especially of other individuals, is mainly a human behavior. However, it depends on what you choose to call "treatment." Candidate activities among birds include dust-bathing, hiding and resting when ill, and "anting"—where they rub their feathers with ants, which then secrete antimicrobial chemicals. Various birds and mammals eat clays to counteract poisons in food, and some types of chimpanzee chew certain pungent leaves when ill. Such "medicines" may control parasitic worms. Since plants and traditions vary by region, those habits clearly get passed on as learned knowledge.

Jon Richfield

Saliva contains a complex cocktail of enzymes, many of which have antibacterial properties. In addition, it contains epithelial growth factors that promote healing in the wound; and the act of licking will tend to debride and remove gross contamination from the affected area. At the same time, of course, saliva contains huge numbers of various bacteria. Fortunately these are largely beneficial or have no effect, and there is no evidence to suggest they are detrimental to wound healing.

D. L. Harris

Peruvian macaws are known to eat clay from riverbanks in behavior known as geophagy. Animals and birds are often observed behaving in this way, but this is usually to provide grits for grinding food in their gizzards or supplying essential minerals to their diet. The macaws, however, consume only one particular type of clay, which is low in biologically relevant minerals and also far too fine to exert crushing and grinding effects on food in the gizzard. Instead, the birds are self-medicating to protect themselves from poisons. The clay is positively charged, and in the birds' stomachs it binds to negatively charged toxic alkaloids that have been ingested from unripe fruit and seeds. This protects the parrots from the effects of the alkaloids, while giving the macaws an ecological advantage over other animals and birds, which cannot consume the same unripe foods.

Patrick Walter

A previous correspondent wrote "saliva contains huge numbers of various bacteria. Fortunately these are largely beneficial or have no effect, and there is no evidence to suggest they are detrimental to wound healing."

On the contrary, there is compelling evidence to suggest that these bacteria (including *Streptococcus* and *Pasteurella* species) can colonize wounds and severely compromise healing. The use of Elizabethan collars in dogs and cats to protect both surgical and traumatic wounds is as a consequence of this detrimental effect, and I would strongly discourage pet owners from allowing their pets to lick their wounds.

An alternative but more likely explanation for the behavior described in the question is that carnivorous animals enjoy the act of licking a wound for the same reason that they enjoy the act of licking a bone: they like the taste.

Mike Farrell
European Specialist in Small Animal Surgery
University of Glasgow, United Kingdom

SHAPELY PEAR

Why are pears pear-shaped and not spherical like apples?
John Griffiths

Apples, pears, medlars, quinces, and the fruits of related plants such as pyracantha are known as pomes. The fleshy part of the pome grows from tissue between the stem and the carpels, the female reproductive parts of the flower.

Plant hormones called auxins, whose distribution is genetically controlled, guide all plant tissue growth and form, including fruit shape. As long as the final product does not put the plant at a disadvantage, selection has no influence on the fruit's appearance.

So if the question is attempting to address what purpose

or function the shape of a pear serves, the answer is probably none. Indeed, some varieties of pear are roughly apple-shaped. As for the ones that are not, it is possible to speculate as to why. Some primeval pears might have benefited from having their fruit hanging out from between the leaves, or perhaps developing a longer neck made pears—which are softer and larger than, say, crab apples—less likely to drop prematurely. Or possibly some ancient seed-dispersing birds or bats were able to carry the necked fruit over longer distances.

Evelyn Pell

It is possible that human pickers select for appearance, which has led to certain shapes and colors predominating. After all, it is useful to be able to tell one type of fruit from another on sight. Perhaps this is also why you won't find many cherry-like plums on sale in shops, common though they are when trees are in fruit. Plums also seem to be selected depending on their redness, regardless of flavor, even though yellow varieties often taste much better.

P. G. Urben

In the wild the shapes of apples and pears can seem almost interchangeable. Home gardeners, whose fruit does not have to fit into a cardboard tray, can take particular delight in natural variation.

Apple varieties can range from flattish to knobbly or lopsided. Of the 345 scanned images of historic watercolors of pears from the collection at the USDA National Agricultural Library in Beltsville, Maryland (see http://www.ars-grin.gov/cor/pwc/pyrus-art-t.html), the very first is of an akarayu or red dragon, as apple-like in shape as a fruit can be, yet it is a pear.

Pears have been grown in Asia for more than fifteen hundred years, and there they are characterized by their roundness as opposed to the classic "pear shape" that we know best from varieties which emerged in France in the seventeenth century. This was the golden age of Western pear development, and arose because Louis XIV loved pears. Today there are more than five thousand varieties of pear in existence.

Asian pears, also known as "sand pears," are generally crisper and juicier than Western varieties but are not, as some think, a cross between an apple and a pear; they all feature that essential graininess that makes a pear a pear.

Toshi Knell

DIZZY DOGS

In hot weather dogs keep cool by panting. If I were to do this I would hyperventilate and exhale too much carbon dioxide. How do dogs avoid the effects of respiratory alkalosis?

Andrew Benton

Each breath taken by a human (or a dog for that matter) consists of a volume of air that enters the lungs and a smaller volume that only gets as far as the passages that lead to them. This is the "dead space," so called because no exchange of oxygen and carbon dioxide occurs in the mouth, pharynx, trachea, or bronchi.

Rapid, shallow breathing can affect just this dead space without hyperventilating the gas-exchange part of the lungs, the alveoli. As air passes through the dead space it produces a cooling effect as moisture lining these passageways evaporates. Dogs, lacking sweat glands, use this method to cool

down. Humans have no need of this because we use sweat to cool our bodies, though we can do it. Try "fluttering" your breathing by taking fast, shallow breaths, at least sixty per minute. You will feel a cooling effect in your mouth, but not the dizziness that can accompany hyperventilation. It's hard work though.

John Davies
Anesthetist

ALL ALONE

Do polar bears get lonely? I'm not being flippant, just attempting to find out why animals such as humans or penguins are gregarious while others, such as polar bears and eagles, live more solitary lives.
Frank Anders

Having a gregarious or solitary nature are species-specific survival strategies adopted by different animals and birds. Big predatory mammals such as polar bears, grizzlies, and tigers isolate themselves from one another to avoid competition with other members of their own species. By spreading out, they also expand their feeding grounds and breeding territories. If fellow species members come into close proximity there can be fierce competition for food, mates, and territory. The same is true with many solitary species of bird, such as eagles and condors.

These animals and birds usually pair up during the breeding season to reproduce, and separate soon after successful mating or when they have raised their young ones. In most cases, raising the young is the sole responsibility of females.

Indeed, males of such species sometimes kill their young to increase their own reproductive success.

Social animals, by contrast, find strength in numbers. Animals such as antelope on the African savannah or penguins in the Antarctic form big colonies, where they huddle together for warmth and to alert each other to a potential predator attack. In a large herd or colony, losses to predators are negligible compared with what they would be if the animals were in isolated groups.

Between the solitary and social extremes are creatures like lions, wild dogs, and wolves, which often hunt in groups and display differing degrees of social interaction and cooperation.

A similar question can be asked about why some plants are gregarious while others are solitary. In one intriguing strategy, called allelopathy, gregarious plants secrete chemicals into the soil to reduce competition from related species that cannot survive the presence of these compounds. As with animals, these strategies have evolved to maximize the plants' chances of survival.

Saikat Basu

Bears and eagles rarely associate with their own kind because individuals need to defend their own feeding territories in which food is often scarce. Polar bears live in an environment where the food resources are too limited to sustain a large community, so it makes sense for them to be the only predator in this particular niche. When food is plentiful, both bears and eagles will gather together with a reasonable degree of amity.

The reverse is true for social animals, including humans. Social animals are often prey for other species, and cluster together for safety against predation—though this is only one of the reasons for group formation. But when food is scarce, individuals may break away from the group to find it.

Whether an animal can feel anything resembling the loneliness humans feel is hard to say. However, highly social animals, such as certain types of parrot, seem to be adversely affected when kept alone. Some parrots will engage in bizarre behaviors and can self-mutilate. Some large parrots will even seem to go "insane" if subjected to long periods of isolation.

On the other hand, certain animals that are by nature solitary hardly appear to be affected at all. Some fish, in particular some types of cichlids, will fight viciously with their own kind if more than one is kept in an aquarium.

Guam rails, a kind of flightless bird, are notoriously intolerant of their own kind, which has obviously made breeding them in captivity very difficult.

So the answer to the question is a qualified yes: some animals will feel "lonely" if they are by nature highly social. However, some will only engage with their own at specific times and in a highly ritualistic fashion, such as when mating or defending their territory.

By e-mail, no name supplied

It depends on the bear and the circumstances. Loneliness is a reaction to deprivation of company when company is appropriate. In the case of polar bears, company usually represents competition or threat, so they do very well by themselves, thank you—unless you happen to be small enough to eat. In certain situations, when food and breeding are not relevant, males will wrestle harmlessly to establish dominance, thereby reducing the risks of dangerous fighting when mating time comes, but that is pretty much that.

Cubs want their mother's company for food, protection, and reassurance, and they want each other's company for socialization, warmth, and play. Females want the company of their cubs, but keep other adults (and cubs) at a distance. Once her cubs mature or die, a mother again becomes a loner until mating time, and then tolerates males only briefly. She has no reason to want any company beyond that.

It is all part of the adaptation to their environment. In zoos, where security and food are no constraint, polar bears often seem happy to have the stimulus of company.

Jon Richfield

6 OUR WEATHER, OUR PLANET, OUR UNIVERSE

THE POWER OF SIX

How can water molecules in a snowflake influence other molecules positioned thousands of molecules away to form the same macroscopic shape and create a six-pointed design? In other words, how do snowflakes get their distinctive and symmetrical six-pointed shape?
Don Jewett

They cannot. There is no spooky power at work here. Ice forms into hexagonal crystals because of the sixfold symmetry of the crystal lattice it forms when it is at or below freezing at a pressure of one atmosphere (a number of other ices form at extreme pressures that dance to different lattice tunes).

The crystal itself does not care about the water molecule far away. It is concerned only with the one that might fit a gap in its lattice, which is how crystals grow. If the molecule is moving too fast, it will not settle into the gap. It is much like the ball on the roulette wheel that bounces here and there until it has lost enough energy to be unable to escape from one of the numbered indentations. Similarly, the water molecules bounce and jostle until one has lost enough energy to settle into the gap in the lattice.

In an environment where heat is slowly being withdrawn, molecules will fill the lattice gaps in an orderly manner and the snow crystals we know and love will grow with that sixfold symmetry.

Differences in temperature and pressure in the microenvironment of each snowflake will create enough variation to ensure that the crystals will not actually be identical, but they can be very similar.

A high-speed freeze will result in a dense mass of interlocked crystals that create ice with no apparent crystalline structure. This is seen in fast-frozen ice cubes, which appear clear. But in a slow-freezing, "constrained" system, such as on a pane of glass or a pond surface, gradual cooling will create the typical hexagonal crystal pattern at the advancing edge of crystallization.

Bill Jackson

Snowflakes begin with a dust particle or similar nucleation site around which water freezes to produce a "seed" ice crystal with six-sided symmetry. Water vapor freezes more quickly at the corners of the hexagonal shape, so arms begin to grow from these corners. The arms can grow at varying rates if the crystal is exposed to different atmospheric conditions. As the snowflake moves in the cloud in which it is forming, it goes through regions of different temperature and humidity, which cause the alternate preferential growth of different crystal faces.

The accumulation of these different phases of growth determines the final complex snowflake shape. Because the snowflake is small, each side of it experiences roughly the same local conditions and so grows in the same manner, resulting in an almost symmetrical snowflake, although of course nothing of this size can be perfectly symmetrical. Since the history of each snowflake is slightly different, the final shape of each one is unique.

Simon Iveson

Simple answer: they don't. Photographers search long and hard to find snowflakes that are close to symmetrical. The vast majority are not. Read the experience of one of the world's greatest snowflake hunters at http://www.its.caltech.edu/~atomic/snowcrystals/myths/myths.htm#perfection.

Colin Dooley

FOG WARNING

On cold mornings I notice that train windows mist up on the inside. Once the outside temperature reaches about 42 or 44°F the windows normally clear. If it is raining, however, the windows mist up at these temperatures or even higher. Why does rain on the outside of the train cause the inside of the windows to mist up, even at higher temperatures?

Klaus Forroby

The windows of the train cool through contact with the cold air outside. Heat conducted through the glass (assuming it is single-glazed) will keep the inside face quite close to the outside temperature.

The inside face of the glass will steam up when its temperature is lower than the "dew point" of the inside air—that is, the temperature to which this air needs to be cooled to become saturated, so that moisture begins to condense out of it. In the case described, the dew point of the air in the train could be just above 44°F.

On a rainy day the passengers' clothing will be wet and their umbrellas will be dripping onto the floor and dampening the upholstery. The heat inside the train will cause the rainwater to evaporate, markedly raising the moisture content of the air. The dew point inside the carriage will therefore be higher. Consequently, more moisture will condense on the window on a rainy day than on a dry one. This effect will be exaggerated if windows are shut, trapping the moist air inside the carriage.

Martin Young

One factor not to be dismissed is that rain wets the outside of the window and, as it evaporates, it cools the glass. If the train is moving, wind increases the evaporation and adds to the cooling. So the window glass will be significantly colder than on a dry day of the same temperature. Now, I wonder if I can use this to cool a bottle of beer?

Spencer Weart

Director, Center for History of Physics

American Institute of Physics, Maryland

AN ILL WIND

When you are feeling nauseous, why does a cold environment make you feel so much better? Contrast this to a warmer,

more humid environment, which makes you feel much, much worse.

Brian Woodhill

There seems to be no clear explanation for this effect, but lots of possible ideas.—Ed.

As any real sufferer from motion sickness can assure you, the effect of coolness is marginal. They can still get very, very sick in the teeth of the most refreshing gales.

The main contributor to motion sickness is conflict between visual signals and signals from the semicircular canals located in the ear and associated with balance. Belowdecks on a ship, assorted stresses aggravate the effects, especially lying claustrophobically in a stuffy cabin, being unable to fix your gaze on stable points of reference, and your head swinging passively at the whim of a pitching bunk. Smells of diesel and vomit, engine rumblings, and nonsufferers chattering on about nauseating antinausea remedies do not help.

Erect on deck, with clean air, an unobstructed view of the horizon, and comparative peace, the most noxious sensory triggers and emotional factors disappear. In particular, avoid leaning against stanchions and use the horizon as a stable reference for orientation. You can then ignore the caprices of the rolling vessel, adapt to the motion, and develop sea legs.

When defying the elements and feeling under control in this way proves to be effective, it is no wonder one associates coolness and fresh breezes with freedom from claustrophobically nauseating stimuli. Staring through rolling portholes or hanging over a chaotically heaving taffrail, while gazing down at chaotically heaving water, doesn't work at all, breeze or no breeze.

Jon Richfield

One reason that people who are nauseous feel worse if they move from a cold to a hot environment may be that exposure to heat induces greater expression of the enzyme heme oxygenase. This is a heat-shock protein that is produced throughout the human body. It breaks down hemoglobin, myoglobin, and cytochromes into iron, biliverdin, and carbon monoxide, and increases this process in response to stressors of any kind.

In hot environments, the concentration of carbon monoxide in exhaled breath may be twenty times higher than when in cold environments. Such high levels of endogenous carbon monoxide may cause not just nausea but also other symptoms of heat stress including vomiting, headache, fatigue, and weakness.

Albert Donnay

Once, while working up a ladder, I managed to bring about an undesired impact between the hammer I was holding in one hand and the thumb of the other. A friend of mine, who has experience in such matters, informed me that there was a simple remedy and promptly stamped on my toe. He was quite right; the thumb gave me no further trouble. His toe-stamp remedy is obviously related to this freezing-your-body cure.

Mark Wallace

I wonder whether this is the wrong question. The right one might be "Why do you feel better when you get to a climate you like?"

I once had a nasty and extended episode of tourist's tummy in Mexico in July 1965, developed at a high, cool, dry observatory site. It was beginning to seem as if the nausea would last forever. But then I got off a plane and stepped into the very warm, very moist air of Acapulco and was immediately cured (and also very hungry).

I'm sure that you will get lots of different answers to the question, illustrating a theorem first proposed by Thomas Gold that states that a theorist can explain any correlation, and its inverse.

Virginia Trimble

WATER BOMB

How big is the biggest possible raindrop?
Michael Leonard

Molecules inside a water drop are pulled equally in all directions by the neighboring molecules. The attraction between molecules means the liquid squeezes itself until it has the lowest surface area possible: a sphere. The outer molecules are held by attractions within the drop and from the molecules beside them, providing the surface tension that allows insects to walk on water.

As a drop grows, this weak surface tension becomes increasingly unable to hold the spherical shape and the drop becomes distorted. This is why raindrops smaller than about two millimeters across remain spherical but, as they get bigger, they start to resemble a hamburger bun, with a flat bottom and a rounded top. Air resistance buffets raindrops as they fall, and water does not have sufficient viscosity to damp these disturbances so falling drops usually break up at a diameter of about five millimeters.

These facts were already known by 1904, courtesy of ingenious experiments designed by the Hungarian-born Nobel physics laureate Philipp Lenard and the American Wilson Bentley, best known for his early photographs of snowflakes. Lenard used a vertical wind tunnel to study raindrops as they fell, while Bentley measured the size of raindrops from the tiny pellets of dough they formed when they landed in a pan of wheat flour.

Experiments on the space shuttle showed that in microgravity—and with no wind to perturb them—drops of water can reach in excess of three centimeters in diameter, though footage shows the spheres quivering like jelly, because the surface tension is too weak to maintain the sphere and the viscosity is insufficient to damp out disturbances.

Mike Follows

The average raindrop that reaches the ground has a diameter of between 1 and 2 millimeters. But raindrops between 8.8 and 10 millimeters were recorded twice a few years ago. The first time was by a research plane that flew over Brazil through cumulus congestus clouds that form over atmospheric regions undergoing convection—they are often created by strong updrafts. The drops are believed to have been formed by water condensing on smoke particles from burning forests below. Large drops were also spotted falling through clean, marine air over the Marshall Islands in Micronesia. They had formed within narrow regions of cloud

where raindrops were able to collide and accumulate. It is unlikely that any of these large raindrops would have reached the ground intact because they readily break up as they fall, thanks to air resistance.

Martin Dodds

In the 1957 film *The Incredible Shrinking Man*, the protagonist, played by Grant Williams, is irradiated by a mysterious cloud which causes him gradually to dwindle in size. Near the end of the film he has shrunk to the size of an insect and is trapped in his own cellar, where a leaky tap threatens to drown him in what are, to him, enormous globules of dripping water.

Yet when the studio's props department built their gigantic faucet and rigged it to drip water, the laws of physics kept all the water drops stubbornly normal sized. When I met the film's director, Jack Arnold, years after the film had been made, he told me that he personally hit upon a method of creating the oversized water drops for this scene. He purchased a hundred boxes of transparent latex condoms and the stagehands were given the job of filling them with water.

With the cameras rolling, Williams posed near the enormous faucet while a concealed stagehand pushed the condoms out of it, one at a time. The effect is quite convincing: each condom took a teardrop shape in midair as it plummeted to the ground, where it burst and released dozens of smaller, genuine droplets. Each discrete "water drop" emerging from the giant faucet was slightly smaller than the shrinking man's head.

Arnold told me that when the film's producer demanded he justify his purchase of a hundred boxes of condoms, Arnold replied: "When we finished shooting the movie, we had one hell of a wrap party."

F. Gwynplaine MacIntyre

DISTRIBUTORY DILEMMA

When great ice fields melt, will any resulting rise in sea level be equal at all latitudes, or will the Earth's rotation make it greater at the equator? If there is a difference, has this been factored in when discussing the effect of melting ice sheets?

By e-mail, no name supplied

The pattern of sea-level change is far from uniform, although the effect of the Earth's rotation is rather small.

If the Earth was covered in ocean, and an extra layer of water was added, then this layer would be about 0.5 percent thicker at the equator than at the poles. This is because the effect of the Earth's rotation makes gravity (the sum of the gravitational attraction of the Earth and the opposing force caused by the Earth's rotation) about 0.5 percent weaker at the equator than at the poles. For a seven-meter (twenty-three-foot) sea-level rise (about what would be expected from the melting of either the Greenland ice sheet or the West Antarctic ice sheet), that would mean a rise at the equator of about 3.5 centimeters (1.5 inches) more than at the poles.

A second effect of rotation comes from the fact that the mass of the melting ice is moving from near the poles to a broader distribution around the whole globe, making it move away from the Earth's rotation axis. This increases the Earth's moment of inertia and so, to conserve angular momentum,

the Earth's rate of spin will slow (like a spinning ice dancer slowing down when they stretch out their arms).

Spreading seven meters' worth of water from the pole right across the whole globe would slow down the Earth by about one part in a million, making the day about 0.1 seconds longer and reducing the force that creates the Earth's equatorial bulge. This force would then relax slightly, partially canceling the above effect. Again, we are left with sea-level differences of a few centimeters (about an inch) between the equator and pole.

However, by far the largest effect would be caused by what is known as self-attraction and loading. The Greenland ice itself produces a gravitational attraction that pulls the ocean toward it. As the ice melts, this attraction decreases, and the ocean relaxes away from Greenland. In addition, the removal of the heavy load of ice from Greenland allows the Earth's surface beneath to bounce back up. This causes mass redistributions in the Earth that offset some of the change in the gravity field caused by the loss of ice mass. The net effect is that the sea level within about a thousand kilometers (six hundred miles) of Greenland actually goes down, but the sea level rises farther away by a little extra.

Researchers are now working on using the pattern of sea-level change as a "fingerprint" to help determine whether changes are a result of melting in Greenland or in Antarctica or elsewhere. Measurements of changes in the gravitational field by the GRACE satellite mission provide our current best estimates of the mass balance of these ice sheets.

Finally, things are complicated by the fact that sea level is not actually level. The permanent currents that make up the ocean circulation lead to slopes in the sea surface, making it depart from a level surface by up to a meter, or one yard. Adding such a large amount of fresh water to the ocean is likely to lead to changes in this circulation and to more complicated patterns of sea-level change.

Climate simulations so far have only considered the global mean sea-level change, together with regional patterns associated with ocean currents. Adding in the effects of self-attraction and loading, which can be calculated separately, is quite a new departure.

Chris Hughes
Proudman Oceanographic Laboratory, Liverpool, United Kingdom

THE DAY THE WORLD STOPPED

How much force would be required to stop the world spinning? If you used, for example, the engines of the space shuttle to do it, how long would it take? And what would be the effect on the planet, in particular the weather and the tides?
Stephen Frost

This is an excellent question for practicing numeracy. All that is needed is some basic mechanics, made a tad more difficult for being rotational. I have rounded up a few of the figures.

The mass of the Earth (M) is 6×10^{24} kilograms and its radius (R) is 6.6×10^{6} meters. Assuming it to be a solid homogeneous sphere, its moment of inertia (J) is given by $0.4 \times M \times R^2$. It works out at 1×10^{38} kg m^2 .

The planet spins once in twenty-four hours (86,400

seconds) so its angular velocity (ω) is 4.16×10^{-3} degrees per second, or, more properly, 7×10^{-5} radians per second.

Earth's angular momentum (h) is the product of the moment of inertia and angular velocity ($J \times \omega$), which gives 7×10^{33} newton meter (Nm) seconds. This is the momentum the shuttle engines will have to counter.

The thrust (F) of the shuttle engines on takeoff is around 4×10^{7} newtons and, if acting tangentially at the surface of the Earth, the torque (T)—or rotational force—about the Earth's center is $F \times R$, which gives 3×10^{14} Nm.

This torque acting over time (t) will change the Earth's angular momentum by an amount $T \times t$. The time needed to reduce it to zero is h/T or 3×10^{19} seconds, or 840 billion years.

This is some sixty times the age of the universe, and by the time the shuttle had done its job there would be no weather or tides worth having. There is one other wrinkle: if the fuel needed comes from the Earth, the planet will get lighter and lighter. The whole of the Earth's mass will be expended as fuel long before the Earth stops spinning.

Hugh Hunt
Senior Lecturer in Engineering
Trinity College Cambridge, United Kingdom

Quite apart from the shuttle engines' puny thrust, there is a fundamental problem with using them at ground level to exert the required torque. A rocket engine works by ejecting mass at high speed in the opposite direction to the required thrust. But at ground level, the material ejected would be rapidly slowed by the atmosphere, which would acquire its momentum in the process. So the initial effect of firing the engine would not only be to slow down the Earth's rotation but also to speed up the rotation of the atmosphere. Eventually, friction between the Earth and its atmosphere would tend to slow the atmosphere and speed up the Earth again, so the net effect would be zero.

To avoid this, the rocket motor would have to be tethered high enough for the material it ejected to escape from the atmosphere. The net effect would be to transfer a little of the Earth's angular momentum to the rocket exhaust.

It is fortunate that stopping the Earth's rotation is no easy task, because without it the Earth would be a much less pleasant place. There would be no day and night as we know it: each point on the Earth would be in permanent sunshine for six months of the year and then darkness for six months.

There would still be tides as the moon continued to orbit the Earth, but instead of two tides each day there would be two tides every twenty-eight days. On the positive side, there would be no hurricanes, because without the Coriolis forces provided by the Earth's rotation, air would flow straight from high pressure to low without any swirling.

Ian Vickers

If each day were to last a year, there would be six months of darkness and cold which would kill most plants and wipe out higher life. The dark side of the Earth would get much colder and the light side hotter.

Without rotation, the dynamics of the atmosphere and oceans would alter profoundly. There could be no hurricanes or revolving weather systems, and ocean currents would change too. Once the Earth had stopped turning, the pull of the moon would very slowly spin it back up

until its rate of rotation matched the moon's orbital period.

Because the equilibrium of the atmosphere depends on many biological and geochemical feedbacks, what happened next would depend upon how much land and ocean was left facing the sun. There may even be limits on day length beyond which a complex ecosystem and atmosphere could not exist.

Don't worry, though. The amount of energy needed to stop the Earth turning is so vast that nothing humans can do will ever stop it.

J. McIntyre

ICE SUMS

I have read reports that rivers in the Himalayas will be starved of water when the glaciers disappear. If the glaciers were neither shrinking nor growing, the amount of water fed into the rivers from the glaciers should be roughly equal to precipitation in the Himalayas over the same period. Assuming that rainfall remains the same when the glaciers are gone, why shouldn't the rivers still receive the same amount of water?

Felix Lim

The water in a river does not depend only on the amount of water that falls as rain in a year, but also on when and how much. Generally, glacier-fed rivers will receive a constant trickle from melting ice in the winter and increased amounts as the temperature rises in spring.

If the glaciers disappear, there will be no slowly melting ice, so the glaciers' regulation of the water flow will cease.

Any rain that falls will instead reach the river rapidly, while in dry spells there will be no water at all. So, instead of having a river all year round with a flow rate that is strongly linked to temperature, you will have a much lower water level most of the time, with higher levels after rain. The average level over the year will indeed stay the same, but in extreme cases it could mean the difference between a constant flow and a dry riverbed with occasional flash floods.

The above compares the case of no glaciers with the case of constant glaciers. In reality, glaciers have been losing mass since the end of the last ice age. Once all the ice has melted there really will be less river water available, so the average level will be lower and more erratic.

Rick

Glaciers act as reservoirs that even out the flow of water, so that a given year's precipitation takes generations to reach the sea. Without glaciers to regulate the flow, a rainy season's precipitation—potentially all of that year's rain—could flood down within months, leaving devastation and eroded canyons for the rest of the year. The glaciers' buffering effect is huge.

Alicia Jones

SHIFTING SANDS

How deep is the sand of the Sahara Desert and what is directly beneath it? The Sahara is subject to considerable wind erosion and its sand is found thousands of miles away. Is it replenished by some means?

Barry Ireland

Not all the Sahara is sandy, but wherever windblown sand settles for any reason, more is likely to collect in the same place. In these places, sand accordingly forms wind-driven dunes, whose behavior is astonishingly complex.

Surface sand seldom piles higher than a couple of hundred yards above underlying earth or bedrock, except by filling ancient valleys or lakes. Such deep sands and spongy sandstone form important groundwater reservoirs. And yes, sand does form and reform constantly as water erosion, frost, and wind-driven particles flake grains off rocks. Conversely, deep moist layers of sand become cemented into sandstone, which in turn may go through the same cycle after millions of years.

Far deeper sand occurs in submarine detritus fans, which build up at the mouths of rivers. Even more intriguingly, the Mediterranean has dried up repeatedly in the last tens of millions of years. Each time that happened, rivers flowing into the basin eroded their fans into massive canyons, which silted up again whenever the sea returned. The bed of the lower reaches of the Nile consists of silt, more or less compacted, miles deep. The river's erosive strength can turn this into an underworld canyon that has dwarfed the Grand Canyon in the past.

Jon Richfield

At more than 9 million square kilometers (3.5 million square miles), the Sahara is the world's largest desert, but it has no uniform topography. About 15 percent of the area is covered by sand dunes, 70 percent is made up of stone deserts of denuded rock and coarse gravel, while the rest is oases and mountain ranges.

Beneath the sand dunes lies rock of varying types. In Algeria and Libya, deposits of oil and gas have also been discovered, although inaccessibility has hampered their exploitation.

The same forces that remove sand from the Sahara are also largely responsible for it being replenished. The wind not only sweeps sand from one area, but causes erosion of rocks in others, which leads to it being replaced.

Ian Smith

SUN SOUNDS

Is it true that if the Earth's atmosphere extended all the way to the sun, then the noise the sun makes would deafen us all? How much sound energy does the sun create? And what other effects would manifest themselves?

Pat Morgan

We and the sun are in each other's atmospheres already, but most of the gas between us is so tenuous that the sound of the solar wind against our magnetosphere cannot compete with our traffic, pop music, and strife. Even if that gas was as dense as our atmosphere at sea level, the loudest solar noises would probably be inaudible infrasonic rumblings, and much weakened after such a journey.

We would have more than noise to worry about, however. The sun would hardly be seen: our atmosphere would be too dense for much light to penetrate if it extended the 150 million kilometers (93 million miles) to the sun. Also, our atmosphere would have a mass of well over 15×10^{30}

metric tons—thousands of times greater than the sun plus all its planets. It would collapse and the resulting blast could probably sterilize any planet within several light-years.

Jon Richfield

A sea-level atmosphere stretching to the sun's photosphere would block observation of the sun, visual or auditory. The huge depth of air would block the sun's radiation: sunrises and sunsets demonstrate this effect on a small scale, as they dim and redden sunlight at dawn and dusk.

As far as sound is concerned, we already have a comparable effect. We do not hear a continual barrage of thunderclaps, even though they are always taking place around the world, because of the limited distance sound travels in the atmosphere. So we would not be deafened by noisy sunspots but we would be frozen by the insulating effect of the wall of air.

The scenario ignores the fact that atmospheres around planets vary in density due to gravity. If the Earth's atmosphere were to spread as far as the sun and have sufficient density to conduct sound, it would probably be solid ice for most of the distance, causing strange gravitational and orbital effects and burying a human observer on the Earth's surface.

Michael Burberry

I know it is not in the spirit of these books to challenge the premise of the questions, but to form an image implied by this query would need some sort of pre-Copernican or geocentric model of the solar system. I cannot quite conceive of an Earth that can drag around an atmosphere which reaches the sun. It is much easier to think of the sun's atmosphere extending all the way to the Earth—which, in a sense, it does.

Actually, the answer may eventually be tested experimentally, but not for a while. As the sun reaches the end of its life, it will become a red giant that will engulf the Earth, and presumably we will be first hit by its upper atmosphere. If life as we know it is to survive it will need to move away from the doomed Earth, but perhaps those who leave last, as well as turning the lights out, could be persuaded to leave a few microphones around just to see what happens.

Bryn Glover

A DROP IN THE OCEAN

If you were to take all the boats in the world out of the sea, how much would the sea level drop?

Richard Higginson

According to Archimedes' principle, a body sinks until it displaces its own weight of water. This is why a fully laden ship settles deeper in the water. Each ton of ship or its cargo displaces a cubic meter (approximately thirty-five cubic feet) of water.

Warships are referred to by their displacement. The combined displacement of the world's military fleets is about seven million tons.

The merchant fleet is much larger, but to complicate matters, merchant ships are referred to in terms of deadweight tons (dwt). This is the mass of cargo that can be loaded onto an empty ship before it risks capsizing. It takes no account of the water displaced by the ship before the

cargo is stowed. The *Knock Nevis*, launched in the 1970s, remains the largest merchant vessel afloat. It can carry a cargo of up to 564,763 dwt and displaces 83,192 tons of water when empty.

The world merchant fleet amounts to 880 million dwt. If the empty fleet displaced the same tonnage of water as the cargo it can carry, then the fully laden fleet would displace 1,760 million tons. Add the displacement of the military fleet and the world's large ships will displace 1,767 million cubic meters of water. When spread over the surface of the oceans (about 360×10^{12} square meters), sea level would rise by a mere five micrometers. The slightly higher density of sea water makes little difference to this figure.

Sea freight increases at a rate of about 3 percent annually, equivalent to emptying the contents of ten thousand Olympic-sized swimming pools into the oceans. But this is, indeed, a drop in the ocean—twenty-five thousand times smaller than the annual rise in sea level.

Mike Follows

FREE-FALLING

What is the greatest height from which one could parachute, and why does such a limit exist?

James Tidey

The chief factor limiting the height is the jump vehicle. No plane carrying humans has gone much higher than 26,000 meters (16 miles), and at that altitude it is going too fast to drop a human. Spacecraft travel higher but are faster still, so the parachutist would need a heat shield to survive reentry.

The only human-carrying vehicles that fill the gap between aircraft and spacecraft are balloons, so the maximum height is that which a balloon can reach. This is 34,668 meters (approximately 21.5 miles), in a record set by U.S. naval officers Victor Prather and Malcolm Ross when they flew from the USS *Antietam* in the Gulf of Mexico on May 4, 1961. However, they didn't jump.

The highest-altitude parachute jump was made by Joseph Kittinger of the U.S. Air Force, who jumped from a balloon at 31,333 meters (19.5 miles) on August 16, 1960. He was in free fall for four minutes thirty-six seconds, reaching an estimated speed of 1,150 kilometers (714 miles) per hour. He opened his parachute at 5,500 meters.

Martin Gregorie

Normally, jumps are made from below 4,200 meters (13,780 feet), a limit set by the risk of anoxia. And in jumps from higher up, the thickening air as the parachutist falls can create problems.

In the lower atmosphere a skydiver accelerates down for about ten seconds until the increasing drag matches his or her weight, which happens at a terminal velocity of about 55 meters (180 feet) per second. As air thickens, terminal velocity decreases. For most free falls, skydivers are decelerating.

Falling through higher, thinner air, you would be traveling faster than the terminal velocity of the lower air when you meet it and the drag force peaks. Effectively, you collide with the atmosphere. During his jump in 1960, Kittinger felt this force as a choking feeling, peaking at about 1.2 *g* at around 23,000 meters (14 miles).

A fall from 75,000 meters (46.5 miles) would give a 3 *g*

impact [that is, three times what would be felt at the Earth's surface] with the atmosphere at about 31,000 meters, which would wear out over twenty seconds or so, when the jump would become an uneventful skydive. A skydiver reentering from low Earth orbit need not suffer much more than 3 *g* if they position their body across the airflow to lengthen the time of impact with the atmosphere, but it would get very hot.

I write as a skydiver of two thousand jumps.

Roger Clifton

Kittinger wore a full pressure suit to protect himself from the low pressure of the stratosphere. However, the main problem with this kind of jump is retaining stability during free fall. Kittinger's equipment included a small stabilizing parachute, but this failed during his first attempt, putting him into a 120-revolutions-per-minute spin as he descended. He lost consciousness, only surviving because his automatic main parachute opened.

The highest parachute jump ever seriously contemplated would have been part of Project Moose. This was an American study, carried out in the early 1960s, into a system which would allow an astronaut to bail out of a spacecraft in low Earth orbit. The space-suited astronaut would wear a parachute on the chest and a folded plastic bag on the back. A pressurized canister would unfold the bag and fill it with polyurethane foam, forming a heat shield. The astronaut would use a handheld rocket to drop out of orbit and begin reentry. Protected from the heat of reentry, the astronaut would wait until slowed by the dense, lower atmosphere before opening the chute and discarding the shield.

Work carried out by General Electric showed that the idea, although outlandish, wasn't impracticable. A prototype heat shield was constructed and samples of the foam were flown on spacecraft. However, neither NASA nor the air force showed much interest.

M. T. Morton
School of Computer Science
University of East Anglia, United Kingdom

The farthest distance one could free-fall to Earth using a protective suit is around 320,000 kilometers (200,000 miles). This is roughly the distance at which the gravities from the Earth and moon are equal. Starting from rest, falling from this height it would take you about twenty-four hundred years to reach the Earth's surface.

Kevin Bastien

Someone falling to Earth from space would arrive at the atmosphere at slightly beneath the Earth's escape velocity of 11,100 meters (36,400 feet) per second. This is why spacecraft have heat shields and enter the atmosphere at a shallow angle to allow them to slow.

To lose kinetic energy, space parachutists falling vertically would need to have perhaps two sacrificial parachutes and a third conventional one. The first would be enormous; it would effectively be a solar sail. Parachutists would have to deploy it thousands of kilometers out and lose speed by tacking in the solar wind. This would be tricky if they were coming from inside Earth's orbit, and they wouldn't want to spend too long in the inner Van Allen radiation belt. The second would be a big upper-atmosphere parachute that would take them from a few hundred kilometers to Kittinger's 30-kilometer (19.5-mile) starting point. Then they can free-fall using a conventional parachute.

We don't yet have the technology to make the first two parachutes. And if anyone traveling this way made any mistakes they would become spectacular meteors.

Adrian Bowyer
Mechanical Engineering Department
University of Bath, United Kingdom

Michel Fournier of France has for many years been planning to skydive from 40,000 meters (25 miles). His latest attempt in May 2008 failed when his helium balloon—intended to transport him into the stratosphere—malfunctioned and departed without him. He insists he will try again.—Ed.

SAND ISLANDS

When walking along a beach in Majorca in late April, I noticed some unusual patterns being cast on the seabed by small floating patches of sand. What causes the patches to form and why do their shadows on the seabed have bright fringes around the edges and around the gaps in the middle?
Tim Pickles

The effect is very specific and perhaps more likely to be seen on the west coast of the British Isles than in Majorca. The sand will be fine, dry, and windblown—the sort that forms sand dunes. Blown by the wind, the sand will fly a few inches above the beach, and some will be trapped on the surface of any water in its path, including a calm sea.

Dry sand takes a few minutes to wet through, and initially there is a "contact angle" between the air/water interface and the surface of the sand grain. The result is that the grain will sit on the surface rather as if it were on a trampoline, with the water surface curving down around each grain. The surface tension associated with this also pulls in adjacent floating grains to form rafts of grains. Around these "sand islands" the meniscus of the water effectively forms a convex cylindrical lens.

If the depth of the pool below is close to the focal length of this lens then the sun's image will be projected as a line around the shadow below, as seen by your questioner. Eventually the sand grains become thoroughly wet and sink.

David Stevenson

The effect in question is a result of surface tension. When the sand settles on the surface of the water it presses it down, so that the level of the water under the sand island is slightly below that of the surrounding water, making the surface curve between the two levels. This has the same effect as a convex lens, focusing sunlight on the bottom to create the bright fringe around the island's shadow.

Another example of the same effect is seen around the surface-dwelling insects commonly called pond skaters, which cast disproportionately large and blotchy shadows with a "silver lining" around them. This is because the light that one would normally expect to land directly below the edge of the skater is refracted away from the skater by the curved water surface, effectively enlarging the shadow.

A. Anderson

INFLATION THEORY

If a balloon was inflated with helium and released into space, what would happen to it?
Kathryn Bergin

If the balloon was inflated inside a spaceship at atmospheric pressure and then released into outer space, where the pressure is close to zero, the balloon would probably explode instantly because of the sudden increase in the pressure difference across the membrane. However, if the membrane was strong enough, then the balloon would expand instead of exploding.

If it was inflated outside the spaceship (or survived being released from inside), then it would move in the same way as any other object of equal mass because there is no air resistance in a vacuum. For instance, experiments have been performed which show that a feather falls at the same speed as a lead weight when released in a vacuum.

Simon Iveson

Assuming the balloon does not pop immediately, its fate depends on where it is released and how fast it is traveling.

There are three cases. If it is released at an altitude where atmospheric drag is negligible and it is moving at a high enough velocity in an appropriate direction, the balloon will continue flying around in space until it is punctured by some small object such as space junk or a meteoroid.

If, on the other hand, while still traveling fast, it falls too low and encounters the Earth's atmosphere, then the balloon will vaporize or burn. But if the balloon is released at a low velocity, just above the atmosphere, it will fall until it meets the atmosphere and slows down. The balloon will then reach an equilibrium height where its buoyancy equals its weight. There it will remain until it is punctured, or until the fabric deteriorates and it pops.

E. T. Kvaalen

The simple answer is that the balloon would behave in exactly the same way as any other object released into space: it would continue to travel along the same trajectory until another force was applied to it. If it were released close to a planet, or similar object, then it would move into orbit around it.

This was what happened to one of the United States' earliest experiments with communication satellites, Echo 1A and Echo 2 (Echo 1A is popularly known as Echo 1, but the original Echo 1 satellite was actually destroyed when its Delta launch vehicle failed on May 13, 1960).

The Echo satellites were both balloons (nicknamed "satelloons" by NASA technicians), 30.5 meters (one hundred feet) in diameter when fully inflated, and constructed from a metallic-coated Mylar skin. The satellites were intended to reflect radio transmissions, particularly intercontinental telephone and TV signals. The Echos' mode of operation was entirely passive: radio waves simply bounced off their shiny surfaces, which, combined with their relatively low orbits (between 1,519 and 1,687 kilometers—between 944 and 1,048 miles—above the Earth), made them clearly visible across the entire globe (see http://www.astronautix.com/craft/echo.htm).

Indeed, the satellite's remarkably high albedo, which made it seem brighter than a first-magnitude star (the brightest star in any given constellation), once caused considerable embarrassment to Arthur C. Clarke. At the time, he and Stanley

Kubrick were discussing ideas for a movie script that would eventually become *2001: A Space Odyssey*. Clarke had just talked the director out of including UFOs in the plot synopsis when both men glanced at the sky over Kubrick's New York apartment and were stunned to spot a classic flying saucer passing serenely overhead. It was Echo 1A.

The Echo balloons were also involved in research into atmospheric density, solar pressures, the dynamics of large spacecraft, and global geometric geodesy. Indeed, the Echo program enabled the Pentagon to pinpoint the precise location of Moscow for the unfortunate purpose of accurately targeting the Soviet capital with its missiles. A third NASA satelloon, PAGEOS, intended purely for geodesic studies, was launched into a polar orbit in 1966.

Their success illustrates that a balloon can be safely inflated in space, although there are some distinctive aspects of operating balloons in space. If the envelope is exposed to direct sunlight, then the balloon will spin because the gas on the sunward side is expanding faster. It would also be subject to increased thrust from the gas escaping on that side: no material is entirely gas-proof when it comes to holding either hydrogen or helium, and the expanding gas would stretch the balloon's skin and make it more permeable as a result.

Given sufficient time, all the gas within the envelope would escape. The balloon would not, however, deflate in the accepted sense: instead it would retain its spherical form even when completely empty. This is because there is no external pressure to force it to collapse, as there is in the Earth's atmosphere.

As was established during the Echo program, because of their large surface area and low mass, space balloons are susceptible to pressure from the stream of charged particles known as the solar wind. In fact, away from Earth's orbit, a balloon would tend to drift on this stream, as a terrestrial counterpart would in the jet stream, and would seem to offer a far simpler alternative to the widely touted solar sail as a low-cost deep-space probe.

Hadrian Jeffs

7 TROUBLESOME TRANSPORT

CRASH COURSE

In the TV program Knight Rider, *lead actor David Hasselhoff drives his talking car, KITT (or Knight Industries Two Thousand), up and onto a moving truck via a ramp projecting from the back onto the road, often at high speed. The same thing happened in the original movie,* The Italian Job, *starring Michael Caine, when Minis were driven into the back of a converted bus. But is this possible? As soon as the car hits the ramp, it would be moving relative to the truck and therefore would only have the length of the inside of the truck in which to brake, probably ending up crashing into the cab. Am I correct?*

Jim Bob Hinks

This stunt is possible. It's all to do with relativity. The bus moves at slightly less than the speed of the Minis. If the Mini approaches at twenty-six miles per hour and the bus is traveling at twenty-five miles per hour, relative speed is one mile per hour. The result is almost the same as driving the car into a garage.

I say "almost" because at the moment of entry the Mini's engine, transmission, and wheels are running far too fast for driving at one mile per hour. So in the transition from road to ramp, the front wheels driving the Mini are suddenly spinning at the wrong relative speed. As the wheels touch the ramp, a similar effect occurs as that produced by revving an engine at standstill and then suddenly letting out the clutch: a lot of wheelspin and burnt rubber. In this case, of course, it is not the wheels trying to burn up the road, but the ramp trying to slow down the wheels.

As soon as the drive wheels touch the ramp, the car will leap forward. A dab of the clutch can stop that, and the driver would quickly have to drop down in gear before driving the car into the bus as if it were a stationary garage, except for a little initial acceleration from the rear wheels.

Terence Hollingworth

At a relative speed of about six miles per hour—about jogging pace—a car can easily stop in a short distance, although there will be plenty of screeching as the fast-spinning wheels come to a halt, which is possibly where the confusion lies. Perhaps it is easier to imagine instead a small aircraft landing on the back of the truck: as long as the difference in speed between the aircraft and the truck was small, there would be no problem landing in the limited space.

Moving walkways in airports offer another perspective. In order to avoid being jerked when we roll our luggage trolley onto the moving walkway, we make sure that we're walking at about the same speed as the walkway. Then as soon as we're on board, we stop pushing. Having made sure that the trolley is moving at just the right speed, there's no

more to do once the wheels of the trolley have stopped turning, and friction takes care of that.

Hugh Hunt

Perhaps the most interesting aspect is the consequence of not getting this stunt quite right.

With a manual transmission, any attempt to drive straight from the road slowly up the ramp would be doomed to failure. The engine would stall as the driven wheels hit the ramp. This would leave a front-wheel-drive car such as a Mini with its front wheels stationary and rear wheels spinning at road speed. Using the brakes to hold the car on the ramp while restarting the engine and engaging first gear would slow the rear wheels and pull the car off the ramp, dumping it onto the road at speed while still in first gear. A rear-wheel-drive car like KITT would fare better, but there would still be the danger of slipping back as the engine stalled.

It is hard to predict what a car with an automatic transmission would do. Normally these cars do not stall, but who can predict what would happen as the driven wheels stopped almost instantaneously from high speed?

Using manual transmission, the truck should be approached with enough relative speed to roll up the ramp after declutching before hitting it. The car should then coast up the ramp with a brief squeal from the tires.

A driver who wanted to increase entry speed for dramatic effect should use a heavy car and a lightweight truck. As the car brakes, the dumped momentum would accelerate the truck. The effective braking distance available for the car would then exceed the internal length of the truck, giving more time for the car to stop.

Trying this with a four-wheel-drive vehicle might be

unwise. Many have differential locks to limit speed differences between front and rear wheels. If the transmission survived, the lock would stop the rear wheels as the front mounted the ramp, or more likely keep the front wheels spinning at road speed. Neither would be conducive to the fine control needed.

Mark A. Jones

I have seen interviews with the production team on *The Italian Job.* The stunt people really did drive the Minis onto a moving bus. The bus driver's cab was reinforced to protect him as the cars ran into the back of it. As each Mini came in it smashed the preceding one further along the bus until all were in. However, the force of the cars slamming into each other and the cab caused the driver to be rammed against the steering wheel, and he had to be cut free.

Susanna Sherwin

All your correspondents providing answers to the problems of the *Italian Job* stunt are concerned with the relative speeds of the wheels and the ramp, and the potential for the adverse impact of almost instantaneous changes in speed.

However, there need be no problem. By using a ramp with surfaces covered in small rollers, like those used in industrial conveyer systems, there would be essentially no impact because the rollers have a relatively small inertia and speed up very quickly. The problem of stopping the vehicle once on the ramp could then be addressed by using friction in the rollers to make them progressively harder to turn the further up the ramp they are. The result would be a smooth slow-down and none of the problems that troubled the three Mini drivers in the movie.

David Sharman

COCONUT CRUISE

How long would it take a coconut to float from the Caribbean to the west coast of Scotland?

BBC Radio 5 Live listener

It's an interesting question, and the jury is still out. The coconut palm seed (*Cocos nucifera*) is the best known of the drift fruits, and it is claimed that viable coconuts have been found as far north as Norway. However, these may have been tossed from ships into the North Sea rather than drifting all the way from the Caribbean. The chances are that a coconut would sink long before reaching Scotland, despite being carried by that "river in the ocean," the Gulf Stream.

There is more chance of finding flotsam lost from the cargo vessels that ply our oceans, however. For example, in 1992 an armada of twenty-nine thousand rubber ducks and other bath toys were spilled overboard during a storm from a container ship as it crossed the Pacific Ocean. Curtis Ebbesmeyer, a retired oceanographer, has been tracking their progress. Now bleached white but still identifiable because of the logo "The First Years" that is stamped on them, each duck has a $100 price on its head, an incentive for beachcombers to report their finds and help scientists develop better models of our oceans.

It is thought that a flotilla of these ducks has reached the Atlantic by navigating their way through the Northwest Passage. The ducks have already proved that flotsam travels up to twice the speed of ocean currents.

To revert back to the original question, using this observation and knowing the varying speeds of the Gulf Stream and North Atlantic Drift, it would take the hypothetical coconut about sixteen months to make the journey from the Caribbean.

Mike Follows

The subject of drift objects has occupied the minds of scientists for over three centuries, and of sailors for much longer, long ago leading to experiments. Surface currents determine the rate of travel of drifting objects. Bottles released in the northern part of the West Indies take fourteen months on average to reach European beaches. The quickest recorded passage was 337 days from Hispaniola to southwest Ireland, a rate of twelve miles per day.

The time needed for an object to float from the Caribbean or South America to Scotland would be longer, probably at least fifteen months. About twenty tropical plant species, including coconut, have fruits or seeds capable of remaining afloat in salt water for this length of time—a few seeds have even germinated after floating across the Atlantic. However, the great majority of coconuts found on European beaches have probably been discarded locally or lost overboard from ships.

Colin McLeod

BOUNCE BACK

When a car crashes and its protective air bags are inflated, where do the air bag covers go to stop them from breaking your nose?

Damien Hadley

Air bag covers are formed from molded plastic and have lines built into them that are much thinner than the rest of the cover. When the air bag inflates, it forces its way through the covers, which fracture along these very thin lines. Obviously, it is important that the cover does not become a projectile, so it also has other thin sections which act as hinges. These hinges ensure that the fractured sections of the cover rotate harmlessly away from the occupant, rather like a pair of barn doors swinging open. These thin hinge lines and fracture lines are often visible, sometimes looking like a large "H" (especially in older or less expensive cars).

Air bag engineers pay particular attention to the design and fixing of any logos or manufacturers' badges that are fitted to the cover in the center of the steering wheel to ensure that the badge remains attached to one of the doors formed by the opening of the cover, instead of coming loose and causing injury.

In the case of side-impact air bags contained within the seat, a similar effect is achieved by providing a weakened seam of stitching in the seat cover, immediately alongside the air bag. This is one good reason not to fit seat covers.

Air bag engineers also have to ensure that the rapidly unfolding envelope of the bag moves straight toward the person it is supposed to protect, rather than across the driver's or passenger's face and chest. This involves very careful analysis of folding patterns, predictive software, and analysis of high-speed photography from test firings. Engineers have also learned a great deal from origami.

Ian Gordon

Air bag technology is one of the most important improvements in car safety, but an improperly used air bag can do more than break your nose. Cover flaps are made of light plastic, ductile at working temperatures and designed to rupture to let the bag through as it expands. Fragments have been known to cause injuries, especially to the face, but these are usually mild. Nearly all air bag injuries result from the victim being too close to the air bag or wearing a seat belt improperly. Some of the worst injuries result from putting a child in an improperly placed seat or carrying them on an adult's lap, which in some countries is now a criminal offense.

Remember that an air bag being deployed is like a bomb going off in your vehicle. If you are too close to the explosion, you may be harmed by impact, blast fragments, or even caustic residues. If you are at least ten inches away from the cover flap, properly seated, with your seat belt taut and correctly placed over your hips and shoulder, air bags are excellent. In this case, air bag injuries should be the least of your worries in an accident.

Jon Richfield

You can see a picture of an already deflated air bag, with the split in the covers clearly visible, at http://upload.wikimedia.org/wikipedia/commons/2/20/Airbag_SEAT_Ibiza.jpg.

Simeon Verzijl

LIKE FALLING OFF . . .

Why is it easy to balance on a moving bike, but almost impossible to stay upright when it stops?

Angela Rouse

First, let's dispel the myth, usually put forward at this point, which suggests gyroscopic effects have something to do with this. I built a bike with a reverse-running rotor attached to the front axle, which canceled out all gyroscopic effects, and it was no more difficult to ride than a normal bike. Some simple sums will confirm that this is the case.

The way we stay upright on a moving bike is by active control through steering. This is why we have to learn to ride a bike. If, as learners, we find ourselves falling over to the left then we learn to steer the bike to the left, which generates forces that tilt us back upright again, thereby putting the wheels back under our center of gravity. Beginners are very wobbly, but as we become expert the corrections become smaller and we can ride in a straight line.

The faster we ride, the smaller the steering adjustment needs to be, simply because the bike moves much farther in a given time. When riding very slowly the steering adjustments required are very large. When completely at rest, active steering can do nothing for us.

A good analogy is to ask, "Why is it easier to hop (or pogo-stick) along a straight path than it is to stand still on the ball of one foot?" The reason is that we use each hop to generate correcting forces and also to put our foot down in a new place that is closer to where we need it to be in order to maintain our balance.

Hugh Hunt
Dynamics and Vibration Research Group, Department of Engineering
University of Cambridge, United Kingdom

A moving bicycle is easier to balance than a stationary one because a successful rider has learned to turn the handlebars, and hence the front wheel, in the direction of lean. This steers the bike so that both wheels come back under the rider's center of gravity; in other words, steering undoes the lean. This is why it is impossible, without a good deal of practice, to ride a bicycle with reverse-geared steering in which the wheel moves in the opposite direction from the handlebars.

How do we know that the oft-quoted gyroscopic effect in relation to this problem has nothing to do with it? Because people have built successful "bicycles" that operate on ice skates rather than wheels, and these can be kept upright using the same techniques.

Hudson Pace

This works because when you are moving, you have a control—the bit we call steering—that is not there when you are stationary. If you are leaning left a bit, you steer left; leaning right, and you steer right. Doing so brings the wheels back under the center of gravity.

The geometry of a conventional bike helps you to do this. Staying upright involves a series of small corrections. It even works when riding on fixed rollers under the wheels, as long as the steered wheel is on a roller that is rotating, and so can provide corrections by "steering" the bicycle to left or right.

Gyroscopic effects seem to add to stability, but you can easily ride with tiny wheels, or with reverse gyroscopes to cancel the effect of the wheels, which means that the gyroscopic effect is not the answer to the wider question of stability. Similarly, moving body weight often contributes to balancing, but is not essential: interestingly, it is easier to learn to ride recumbent bikes if you avoid using body weight for corrections.

Jonathan Woolrich

I was pleased to see that the importance of the gyroscopic effect on the stability of conventional bicycles has been downgraded in previous answers, but was surprised that the most important design aspect that makes bicycles easy to ride has been overlooked. Every cyclist knows that above a certain speed it is possible to ride hands-free. Most will also be aware that once a bicycle without a rider has been given a push to a little over walking speed it will stay upright for quite a while. This is because a conventional bicycle is designed to be inherently stable.

This stability is mainly due to the trail. Trail is the distance between the point of contact of the front wheel with the road and the point of intersection of the steering axis with the road. This comes from the built-in tilt of the steering axis and the rake of the front forks. The effect of trail is similar to the action of casters. If the bicycle leans to the left, the contact force at the road will turn the front wheel to the left. This allows the hands-free rider to achieve a degree of steering control by leaning slightly one way or the other. Dynamic analysis shows that trail, together with a gyroscopic effect, can produce stability above a critical speed, trail being the most important factor.

H. R. Harrison

The usual notion is that to steer to the left you push the handlebars to the left. However, experiment by holding the handlebars with your fingertips and you find that pushing the handlebars to the left by a couple of centimeters makes the bike turn right, rather than left. This counterintuitive effect arises because turning the handlebars a little to the left makes the bike lean to the right, which then "turns" the bike to the right.

Mark Pettigrew

For those who would like to see a bicycle that cancels out gyroscopic effects, Hugh Hunt of the Department of Engineering at the University of Cambridge has posted some images (http://www2.eng.cam.ac.uk/~hemh/gyrobike.htm). In 1987, New Scientist *reported the work of Tony Doyle, then at the University of Sheffield, United Kingdom, who built a bike that not only canceled out gyroscopic effects but also had no trail, and so no caster effect (April 30, 1987, p. 36). "Once [riders] had overcome their initial impulse to scream, they could ride the destabilised bike easily," the article said. "But whereas a normal bicycle stabilises itself almost instantly, when the riders were left to make the corrective movements for themselves, they could do so only after a delay." Doyle also describes the sequence of events needed to turn a bicycle: to begin turning right when traveling at a fair speed, cyclists do indeed push the handlebars to the left, and continue doing so throughout the turn.—Ed.*

SLEEPING SATELLITES

A favorite pastime of ours while we were on holiday in Spain was to gaze at the night sky. There were no city lights nearby, no clouds, and no moon to illuminate the sky. This meant that satellites roaming the night sky were a very common sight. However, occasionally a satellite would illuminate brightly and then fade. What caused this?

Jan Krokowski

This usually occurs around an hour before sunrise or after sunset, when the sun is not visible from the ground but is still visible from the low Earth orbit of many satellites.

Sunlight reflects off a surface on the satellite, such as a

solar panel, causing the bright flash of light visible from the ground. The light fades quickly as the satellite moves on and no longer reflects the sunlight to the observer. Effectively, the satellite acts as a moving mirror, tracing a path of reflected sunlight along the ground.

Kin Yan Chew

In the 1960s I operated the Royal Observatory Edinburgh's satellite kinetheodolite—a device for tracking airborne or orbiting objects—and the Smithsonian Astrophysical Observatory's twenty-inch aperture f/1 Baker-Nunn satellite-tracking Schmidt cameras.

Of the hundreds of satellites which I observed and photographed, many changed in brightness as they crossed the sky. This is caused by the changing aspect of the satellite during its orbit, as seen by a fixed observer, and in some cases its tumbling or rotating motion, presenting different faces that reflect sunlight. The effect is particularly pronounced with those satellites that have large flat surfaces. Satellites also disappear into the Earth's shadow in the evening (for a satellite moving west to east) and appear out of the shadow in the morning.

Today I am a member of the local astronomy society in Guernsey. With many thousands of satellites now in Earth orbit, I typically see about thirty on clear nights, many changing in brightness. Of special interest are the Iridium telecommunications satellites, whose highly reflective antennas produce extremely bright, directional flares lasting a few seconds, some of which can be seen in daylight.

Predictions of when you can see these and other satellites can be found online (see http://www.heavens-above.com).

David Le Conte

In the mid-1960s I worked on a communications project known as Space Junk, which involved bouncing microwave signals off the growing numbers of satellites and their final-stage rocket bodies then in low Earth orbit. In order to accurately point the dishes at the satellites at night, we used an optical tracker to augment and improve the theoretical predictions. Consequently I became very familiar with the visual characteristics of these various pieces of junk.

We observed that while some objects had a fairly constant brightness, others varied strongly and some exhibited a regular flashing pattern. We found that there was a precise correlation between the type of object and the observed visual signature.

The actual satellites, which in general had a fairly regular shape, had a relatively constant brightness, while the rocket bodies, which generally are cylindrical, exhibited a flashing pattern because they are not stabilized and usually finish up tumbling end over end. The tumbling rate can vary widely from seconds to minutes.

These days there are thousands of visible objects in low Earth orbit and each one will have its own visual pattern.

Richard Harris

One of my "things to do before I die" includes making an Iridium telephone call at the same time as an Iridium satellite flare is overhead. There would be something poetic about seeing the satellite being used to relay my call while whispering sweet nothings to my fiancée. She, however, prefers flowers.

Barry Hahn

Your previous correspondent writes that he would like to simultaneously hear from and visually observe an artificial satellite. This is certainly possible if you catch the correct pass of the International Space Station and listen on its VHF amateur radio downlink frequency (see websites below).

In 1991 I organized an amateur radio link between British astronaut Helen Sharman's Juno mission on the Mir space station and a number of British schools. A week before the contact in May I made a test transmission to cosmonaut Musa Manarov on board Mir. As I spoke with him, Mir was clearly visible in its twelve-minute transition across a clear night sky.

A few years later I was asked to give a talk on the amateur radio aspects of the Juno mission to the Denby Dale Amateur Radio Society. There happened to be a visible pass that evening, so at the appointed time the members were ushered outside to observe Mir and simultaneously listen to the cosmonauts' transmissions. This pass was made even more special by the sight of a supply craft that could be clearly seen a short distance behind Mir and which was due to dock a few hours afterward.

Information on amateur radio and how to listen for the ISS may be found at http://www.rsgb.org.uk and http://www.uk.amsat.org.

Richard Horton
Director of Applied Physics and ICT Systems
Harrogate Ladies' College, United Kingdom

NASA's Web page outlining how to use amateur radio to listen to the International Space Station may be found at http://spaceflight.nasa.gov/station/reference/radio/.—Ed.

WIND OF CHANGE

I know that it's bad for fuel efficiency and the environment if I turn on my car's air-conditioning on a summer's day. But surely it is also bad to drive with the windows open because as my speed increases so does the effective ventilation and the fuel cost, because of the added aerodynamic drag. So at what speed is it most environmentally friendly to simply roll up the windows and turn on the air-conditioning?

Duncan Simpson

This is one of those questions that may have no straightforward answer, simply because so many factors are involved.—Ed.

There have been a number of studies on the efficiency of cars in relation to open windows and air-conditioning. In 1986, Cecil Adams of the syndicated newspaper column "The Straight Dope" attempted to test this (http://www.straightdope.com/classics/a2_393.html). Driving a four-door Pontiac 6000LE, he drove three hundred miles at an average speed of sixty miles per hour.

His results showed that he could achieve 35 miles per gallon with the air-conditioning off and the windows up, 34.4 miles per gallon with the windows up and air-conditioning on, and 33.8 miles per gallon with four windows down and the air-conditioning off, showing it to be more fuel efficient to have the air conditioning on and the windows up. This study was conducted in Ohio in May.

An investigation by the Florida Solar Energy Center reached a different conclusion. Conducted with a Volkswagen GTI, this test found that at sixty-seven miles per hour the increase in consumption was 3 percent with the windows

down, whereas with the air-conditioning on it was 12 percent. This study was conducted in July in Florida. I'll leave it to somebody else to sort out if the month, geography, and climate are affecting the results.

David Liptrot

It will vary from car to car, depending on how aerodynamic the car is and how heavy it is.

Another factor to consider is the difference between the ambient temperature and the temperature selected on the air-conditioning controls. In short, it is not possible to quote one speed that is accurate for all cars.

Peter Sharpe

The TV show *MythBusters* tested this using identical cars with the same amount of fuel. It found that at speeds of up to fifty miles per hour it was more efficient to leave the windows open, and at higher speeds it was better to use air-conditioning.

Hoff Wendell

A study in 2004 for the U.S. Society of Automobile Engineers shows just how difficult this question is to answer (http://www.sae.org/events/aars/presentations/2004-hill.pdf). It looked at the impact of air-conditioning on the fuel efficiency of a large family car and an SUV. Tests carried out at General Motors' desert proving ground in Mesa, Arizona, showed that at speeds above thirty-five miles per hour winding down the windows is preferable to switching the air-conditioning on.

The results need to be interpreted with caution, however, as they were sensitive to a variety of factors including external wind speed and direction, and ambient temperature. Unsurprisingly, fuel economy at high speed is much better in more streamlined cars.

Lionel Cooper

CHAIN GANG

My father used to hang a chain which dragged along the road from the back of our car. He said it would prevent my sister from getting car sick. I thought it was some kind of placebo effect but later I discovered that my husband's family did this for their carsick dog. Does it work and, if so, how?

Ginette Andress

The use of chains to cure car sickness is an example of pseudoscience based on a fundamental misapprehension of the real cause of the complaint.

Car sickness, like other forms of motion sickness, is caused by a conflict between the perception of movement by the eyes and that sensed by the vestibular system of the inner ear. In the case of car sickness, the nausea can be exacerbated by feelings of claustrophobia, unpleasant smells, overeating, or lack of ventilation—all classic causes of nausea in their own right, of course (see page 159).

However, it was once widely believed that the cause of car sickness was undischarged static electricity generated by the friction between the car's tires and the road. A car is essentially a metal box insulated from the ground by its rubber tires, which is why you are generally safe from a lightning strike if you're sitting in one with the doors closed. To put it technically, the car body acts as a Faraday cage. This led people to

believe that car sickness might be relieved by discharging the static via a conductive chain linking the chassis to the ground.

Static discharge chains have long been used by tankers delivering gas. Any residual charge in the vehicle could create a spark when its delivery hose was brought close to the filling station's storage tank, so it became a safety requirement that the chain be lowered to eliminate the charge before any fuel was pumped into the tanks. The theoretical risk of an electrical discharge causing an explosion at gas pumps is also the reason that mobile phone use is banned at some filling stations.

The chains on tankers, however, were far more substantial affairs than the flimsy aluminum-and-rubber versions sported by many cars between the 1960s and the 1980s. Even if there had been any validity to the theory (except in terms of profit for the suppliers who sold them to gullible motorists), any beneficial effect would have been mitigated as soon as the car built up speed, and the chains were inevitably lifted off the ground by the slipstream to leave them waggling forlornly in midair.

Hadrian Jeffs

8 BEST OF THE REST

FLIGHT OF FANCY

A truck driver approaches a bridge that has a weight limit of 10,000 pounds. He and his truck weigh 9,950 pounds so he would be able to cross it were it not for his 100-pound cargo—a flock of pigeons loose in the back of the truck. He has the bright idea of banging on the side of the truck to scare all the birds into taking flight and then he quickly drives across the bridge. Does it work?

David Thomas

Unfortunately, it is likely that the pigeons, the truck, and driver would perish as the bridge collapsed. Instead of birds, imagine the truck was carrying a large container within which was a helicopter in flight (setting aside the troublesome flight dynamics involved). The truck would still be supporting the weight of the helicopter via the downdraft from its rotors bouncing off the floor. The same will be true of the downdraft from the pigeons' wings.

Leigh Hunt

The answer depends on the design of the cargo hold. If it is a sealed box, then the plan will fail. When the birds are roosting, their weight is transferred directly to the truck frame. If they are flying, then their wings generate a circulation of air that exerts an excess pressure on the bottom of the cargo hold, and this pressure conveys the same force as the direct weight of the birds. At steady state the entire contents of the sealed cargo hold can in effect be considered as a box of fixed weight. The exact distribution or activities of the individual masses within this volume are irrelevant.

The only possible way out if the birds are in a sealed cargo hold would be to employ a non-steady-state solution. If the driver could train the birds to all stop flapping their wings at the moment when the truck crossed the bridge, then the birds would go into free fall and head downward under gravity, not exerting any force on the truck until they landed on the floor.

The timing would be tricky though. A 6.5-foot-high cargo hold would only afford 0.64 seconds before the birds fell from top to bottom under gravity. You could double the time available if you could train all the birds to launch themselves from the base with just enough momentum to get them up to the roof and then back down without flapping at any stage. In effect this would be like a juggler hurling the birds into the air, crossing the bridge, and then catching them again on the other side. It would, of course, be difficult to make this work in practice.

A more plausible solution would be available if the cargo hold had large openings on the top and bottom, or better still if its roof and floor were made entirely of an open wire mesh. Air would then be able to flow freely in at the top and out at the bottom, so the pigeons' weight would be supported by the flowing airstream and not transferred to the truck. This airflow would generate a small downward shear force on the vertical walls of the cargo hold and on the wire mesh, but this force would be much smaller than the static weight of the birds and thus enable the truck to cross successfully.

However, this solution would also require holes in the bridge to allow the air to flow down through it so that the birds did not impose a simultaneous downward force on the road surface. Ideally the bridge would be just two planks of wood for the truck tires with a gap in between.

Simon Iveson

The practical engineer's answer is yes, of course the driver could cross. The surplus weight of fifty pounds translates to a less than 1 percent excess over the bridge's specified load maximum. An additional impulse force due to the truck bumping over a small stone in the road would be much greater than this 1 percent, and any civil engineer who designs a structure of any sort with a safety margin anywhere near as small as 1 percent deserves all the professional liability lawsuits he or she gets.

Franc Buxton

LIGHTBOX

If you have a hollow cube whose internal sides are made of perfectly reflecting mirrors, and you switch on a flashlight and wave it around, then turn it off, would the light keep bouncing around the cube or would it go dark? If it went dark, where

would the light have gone? If the light continued to bounce around, how long would it do so? This question has perplexed me since I was a boy.

Paul Harwood

If the cube was made out of perfect mirrors then yes, the light would bounce around forever. Unfortunately, mirrors are not perfect—some of the light that falls on them is absorbed. A domestic mirror reflects only about 80 percent of the light falling on it. If you stand between two large mirrors, set up so you can see the series of reflections, you find they soon get noticeably darker. Even a high-quality telescope mirror only reflects between 95 and 99 percent of the light.

The other factor to consider is the speed of light. In a one-meter cube made of mirrors with 95 percent reflectivity, light would be reflected three hundred times in a millionth of a second and lose 5 percent of its brightness each time, so reducing it to under a millionth of its original brightness.

So where does the light go? As the light is absorbed it warms up the surface that absorbs it, so the cube would be ever so slightly warmer.

John Romer

The answer is yes, if the mirrors are perfect and there is absolutely nothing inside the box, including air. Unfortunately, common mirrors are imperfect. After a large number of bounces, which occur quickly given the speed of light, the light would be almost completely absorbed by the mirror. Light would also be absorbed by the air within the cube, but to a much lesser extent.

Perfect mirrors do exist, relying on the principle of total internal reflection. For example, light traveling from water to air can only escape at steep angles. At shallower angles, the light is perfectly reflected back into the water from the underside of the surface. This can be easily seen in an aquarium or even a glass of water. Look at the surface of the water from a shallow angle underneath and you will observe an image due to total internal reflection. Interestingly, in this case the mirror doing the reflecting is neither the water nor the air, but rather the air/water interface.

In order for total internal reflection to occur, the material in which the light is traveling must have a higher refractive index than the adjoining material. However, despite the reflection being perfect at the interface between the two, the first material will absorb some of the light as it travels through. So, unfortunately, there are no perfect mirror boxes—but you can come close.

A fiber-optic cable uses total internal reflection to allow light to travel along it with very little loss over long distances. Such a cable is analogous to a mirror box with two ends extremely far apart. And diamonds are cut to take advantage of total internal reflection, so light will bounce around inside them many times before escaping pretty much undiminished. This gives diamonds their brilliance.

Physicists have created a "mirror box" using a sphere. In a high-Q microsphere resonator, light is trapped in a tiny glass sphere, continuously bouncing off the inside surface at shallow angles, thanks to total internal reflection. Light is continuously pumped into the sphere, and because no light can escape and only very little is absorbed by the glass, the light inside the sphere builds to very large intensities.

These microspheres are used as very sensitive sensors,

detecting impurities that land on their surface because of the way the impurities disrupt the total internal reflection.

Quinn Smithwick

LOUD SHIRT

I recently appeared on my local TV station. When I arrived at the studio I was asked to change my shirt because its pattern would look distorted on-screen. What causes this effect and why, in this day and age, is it impossible for television to simply record what is in front of the cameras?

Alan Francis

There are two possibilities. First, analog television in the United Kingdom uses PAL [phase alternating line] color coding, which transmits the color information as very fine luminance patterning on a simple black-and-white signal. If you feed a color TV signal into a high-resolution, black-and-white monitor, strong blues and yellows appear as fine backward-slanting stripes and strong reds and greens as forward-slanting stripes. Mixed colors, such as orange or cyan, make cross-hatch patterns. In addition, a color TV receiver can mistakenly decode fine-patterned shirts as encoded colors that flicker and are generally annoying. The decoders in newer TVs are much better than those of the 1960s, when the system was invented, but it is a fact of the coding standards that color and luminance cannot be fully separated under all circumstances.

The second possibility relates to the bandwidth of the broadcast signal. Studio-grade digital television operates at a data rate of 270 megabits per second, but for digital broadcast by satellite this signal is compressed to between 2 and 5 megabits per second. To achieve this, many of the twenty-five frames that are shown per second are not transmitted in their entirety; instead, the differences between one frame and the next are sent. Busy patterns on clothing can cause stress in the part of the coder that monitors changes from one frame to the next, resulting in shimmering of the cloth and degradation of the entire picture because too much data capacity has been used up on the complicated cloth pattern.

Broadcasters now transmit four or five digital TV stations in the channel previously occupied by a single analog program, so it is hardly a surprise that something gets lost along the way.

A couple of images on my website show luminance/chrominance coding patterns (see http://www.techmind.org/vd/paldec.html). You can also find the BBC's research and development group's white paper on digital TV compression issues at http://www.bbc.co.uk/rd/pubs/whp/whp131.shtml.

Andrew Steer

There are several reasons why your shirt may be unsuitable for TV. If it has a complex pattern, it may shimmer annoyingly because of the phenomenon of "aliasing." To understand this more easily, imagine stripes that, when appearing on the screen, are closer together than half the spacing of the screen pixels.

Obviously the TV cannot resolve these stripes because it doesn't have the resolution. The problem can be reduced by higher-resolution TV but will never be completely eliminated. This is the reason why wagon wheels appear to go backward in movies.

Also, certain very bright colors, known as "hot colors," can fall outside the range of colors that can be encoded by the TV signal. If your shirt is extremely loud, this could be the problem.

Finally, if the TV studio was using a blue or green screen behind you and your shirt contained a color similar to the screen, this would have the effect of making your body transparent, which would be embarrassing.

Jerry Huxtable

An earlier answer to this question suggested that wagon wheels appear to go backward in movies and on TV due to the finite horizontal resolution of the camera/TV system. In fact, this illusion is caused by the strobing effect of the frame rate. Essentially, each frame (twenty-five or thirty per second, depending on the country) takes a snapshot of the position of the spokes. If the rate of movement of the wheel is such that the next spoke in turn hasn't reached the position of the previous spoke by the time the next frame is taken, the brain will perceive a reverse rotation.

Max Dirnberger
Senior Electronics Engineer

LUCKY NUMBERS

To enter the national lotteries of many countries, you select six numbers. I always feel that I must choose at random, yet the odds would be the same if I always chose the numbers 1 to 6. Why do I feel this way?

Alice Pearson

As you acknowledge, the set of six numbers 1, 2, 3, 4, 5, 6 is neither more nor less likely to win than any other, but while the machine selecting the winning numbers does not prefer particular combinations, the same is not true of humans. We have brains adapted to make sense of the world by looking for patterns, and for us a set such as 15, 18, 23, 31, 37, 49 does not seem to possess any significant pattern to distinguish it from the numbers selected on other weeks.

Our brains cope with a large number of individuals by stripping them of their individuality, so we regard all sets of numbers which lack an obvious pattern as being essentially similar, and do not differentiate between them. Any set that does possess an obvious pattern feels like an unlikely candidate for winning because we perceive it as distinctive, special, and—by perverse logic—unlikely to be drawn.

Despite the fact that 1, 2, 3, 4, 5, 6 is no less likely to win than any other set of numbers, if you're really hoping to win big money, this might not be the best set to choose. I have seen estimates that if all six were ever to come up together, the jackpot would have to be shared between as many as ten thousand people because these people have chosen these numbers after also seeing through the meaninglessness of the patterns we like to impose on random numbers. What they have not realized is that there are a large number of like-minded people with whom they will have to share their winnings.

In this respect, your instinct to choose numbers "at random" may yet have some rational justification. Probably the best tactic is to choose numbers that other people shun—there are a number of websites that offer advice on which numbers are more or less likely to appear and in what months

or weeks of the year. You can choose to take the advice or ignore it, bearing in mind the draw is completely random.

Interestingly, selecting 1, 2, 3, 4, 5, 6 for the British Lotto draw on the date of the week I wrote this in February 2007 would have won £10 for three matching numbers—the first time this set would have won anything for at least the past fifty draws.

Stephe Ellis

There are around fourteen million ways of selecting six numbers from forty-nine, and although each of these ways has the same probability of occurring, there are far more sequences that look random, such as 4, 16, 27, 35, 48, 49, than ordered ones like 1, 2, 3, 4, 5, 6. So, if you had to bet on a random sequence or an ordered sequence being drawn, then the money is with the random sequences every time.

Steven Winfield

COUNTER-TASERISM

At a recent conference of the Police Federation of England and Wales, there was a call for the increased use of Tasers (and, of interest to etymologists, the appearance of the verb "to tase"). Purely in the interests of research, what could be done by an individual to lessen the effect of being tased? What happens if the tased person grabs the officer who has fired the Taser? If the tasee is wearing rubber-soled shoes, will that help? Alternatively, if they are standing in a puddle, will that worsen the effect? And what would a master criminal need to become invulnerable to tasing—a full-body rubber suit?

Adil Hussain

Tasers work by shooting two darts into the skin of the subject and running high-voltage, low-current electricity through the muscles, resulting in painful spasms and loss of voluntary control. Wearing a pair of rubber shoes or standing in water would have no effect because the current passes between the two darts in the skin and not through your body to the ground.

An all-encompassing rubber suit would only protect you if it was thick enough to prevent the darts from penetrating through to the skin. Thick clothes, too, could have this effect, but because a Taser will still be reasonably effective even if it strikes an extremity like an arm or a leg, any layer would have to cover your whole body to be effective, which would be hot and cumbersome.

The best defense is probably to keep moving as fast as possible across the assailant's arc of fire, making it as difficult as possible for the Taser operator to keep you in their sights and therefore maximizing the chance of one or both of the darts missing your body.

Peter Oliver

The best way to stop a shock from a Taser would be to wear a conductive suit in order to short-circuit the needles. A shark suit, which is made of metal mesh and worn by divers in shark-infested waters, might be the most effective. The suit would probably stop the needles reaching the skin, but could be a bit heavy and awkward for a master criminal attempting a quick getaway.

Aluminum foil is cheap and should work, but could pull away from the needles under the force of their impact, so the best option would probably be to wear a woven metal fabric, which would tend to grab the needles.

A bulletproof jacket might also work by stopping the needles before they could penetrate the skin.

Andrew Hicks

Here on the west coast of Canada we read almost daily in the newspapers of people who have successfully lessened the Taser effect. Strangely, they are always found to be high on the drug crystal meth. It seems that they can fight on through half a dozen zaps. Of course they are not sane when the process commences and neither are they when it finishes. One thing is for certain, you shouldn't try this at home.

Jon Ackroyd

The item below was doing the rounds of the world's e-mails in early 2008. It appeared first in a college alumni newsletter in the United States and describes an individual's detailed account of his own experience of self-tasering. It may be apocryphal but we suspect not. Either way we include a précis of it here as a salutary warning to all.—Ed.

Last weekend I came across a 100,000-volt, pocket-sized Taser for sale as a means of self-defense. The effects of the Taser were supposed to be short-lived, with no long-term adverse effects on your potential assailant.

Long story short, I purchased the device and brought it home. I loaded two AAA batteries in the darn thing and pushed the button. Nothing—I was disappointed. I quickly learned, however, that if I pushed the button and pressed it against a metal surface at the same time I'd get a blue arc of electricity darting back and forth between the prongs.

Unfortunately, I have yet to explain to my wife what that burn spot is on the face of the microwave.

So, I was home alone with this new toy, thinking to myself that it couldn't be all that dangerous with only two AAA batteries in it. There I sat in a pair of shorts and a tank top with my reading glasses perched delicately on the bridge of my nose, instructions in one hand, and Taser in the other, thinking that I really needed to try this thing out on a real target.

The directions said that a one-second burst would shock and disorient your assailant; a two-second burst was supposed to cause muscle spasms and a major loss of bodily control; a three-second burst would reportedly make your assailant flop on the ground like a fish out of water. Any burst longer than three seconds would be wasting the batteries. And all the while I'm looking at this little device loaded with two tiny AAA batteries and thinking to myself: "No possible way."

What happened next is almost beyond description, but I'll do my best. I'm sitting there with my cat looking on with her head cocked to one side as if to say, "Don't do it," reasoning that a one-second burst from such a tiny little thing couldn't hurt all that bad. I decided to try just for the heck of it. I gingerly touched the prongs to my naked thigh, pushed the button and . . .

I'm pretty sure Hulk Hogan ran in through the side door, picked me up in my chair, then body-slammed us both on the carpet, over and over and over again. I vaguely recall waking up on my side in the fetal position, with tears in my eyes, body soaking wet, both nipples on fire, testicles nowhere to be found, with my left arm tucked under my

body in the oddest position, and tingling in my legs. It hurt like hell.

A minute or so later (I can't be sure, because time was a relative thing at that point), I collected my wits (what little I had left), sat up, and surveyed the landscape. My bent reading glasses were on the mantel of the fireplace. How did they get up there? My triceps, right thigh, and both nipples were still twitching. My face felt like it had been shot up with Novocain, and my bottom lip weighed a ton. I'm still looking for my testicles. Do not try this at home.

DIVE, DIVE!

In films, a hero often evades bullets by jumping into a river or lake. How far below the surface do they need to dive?
Christian Dawson

Any object moving through a medium experiences a drag force tending to slow it down. For a denser medium like water, the drag force is much larger than it is in air. Water is seven hundred times denser than air. The drag force on the bullet scales as the square of the velocity and is also proportional to the surface area of the moving body.

Knowing this, one can set up an equation of motion for the bullet, which gives the distance over which its velocity is considerably reduced. The formula involves the velocity, mass, and size of bullet, the density of water, and the drag coefficient.

For a typical bullet with a velocity of a thousand feet (three hundred meters) per second, the depth over which it slows in water is barely ten to twelve feet (a few meters). So a ten-foot (three-meter) dive below the surface is more than adequate.

C. Sivaram

Indian Institute of Astrophysics, Bangalore, India

If the villains are standing on the bank, the hero need only be a centimeter or two below the surface because any small-arms bullet aimed at them will glance off the surface just like a skimmed or skipped stone.

If the villains are shooting from an aircraft, the bullet enters the water at a steeper angle. Even then, a .50 armor-piercing bullet will penetrate only a little more than thirty centimeters and a .303 full-jacketed, sharp-nose military bullet only five centimeters. Pistol bullets have rounded noses and will not even penetrate that far.

These figures were obtained by the U.S. Bureau of Ordnance after the Second World War to determine whether water provides protection from machine-gun fire.

So it's true. Swimming underwater is a good way to evade bullets.

Ross Firestone

We had lots of answers mentioning the TV show MythBusters. *Here is one of the best. Check out http://dsc.discovery.com/fansites/mythbusters/episode/episode_03.html for more.—Ed.*

To test this, *MythBusters* created a rig in a swimming pool that allowed the depth of the target to be adjusted. The target was a lump of ballistics gel, which has similar characteristics to a human body. The weapons were fired vertically

into the water from about 120 centimeters above the surface.

The presenters used two small-caliber guns and three high-powered rifles. The small-caliber guns managed to penetrate the ballistics gel after passing through up to 2.5 meters of water; in deeper water, the bullets just bounced off. The three high-powered weapons fared much worse. They fire rounds at supersonic speeds and, although they made a terrific boom as they entered the water, the deceleration forces of this impact caused them to shatter within a few centimeters of the surface, and they failed to penetrate the gel.

So *MythBusters* proved you would be safe from any gun if you dived more than 2.5 meters below the surface.

Michael Aydeniz

There is another important matter to consider. Refraction of light at the surface is more pronounced at shallow angles, and would protect you both by breaking up your image—making it harder to see you—and by causing attackers to aim in the wrong place.

Refraction will make it seem as though you are closer to the surface than you actually are. Even if you are clearly visible, shooters might have to aim a meter or more below where you seem to be in order to actually hit you.

Hollywood has overlooked that there is a much more effective way to dispose of someone sheltering under water: throw in a couple of grenades. Water transmits shock waves from an exploding grenade much more effectively than air because it is less compressible.

Phil Stracchino

Our hero might not be so lucky if he or she were being shot at with supercavitating bullets. Ordinary cavitation occurs when water is forced to move at high speed: when forced past a rapidly spinning propeller, for example. According to Bernoulli's principle, the faster a fluid moves, the lower its pressure. If the pressure falls below its vapor pressure, the liquid evaporates and creates bubbles or cavities.

Under normal circumstances, these bubbles soon implode, but when they envelop the whole object that created them the more stable phenomenon of supercavitation takes over, since there is no longer a point on the object where the bubble edge touches it, meaning it is less likely to implode. Bullets travel faster and farther if they "fly" within their own bubble of gas. To achieve this, they need to be flat-nosed, because this shape forces water sideways at high speed.

Supercavitating bullets are being developed by the U.S. Navy's Rapid Airborne Mine Clearance System. It is claimed that they have detonated mines fifteen feet (five meters) below the water surface.

Mike Follows

Water acts as a very efficient absorber of energy, something that has been well understood by navies for centuries. Tests showed that a round shot, or cannonball, would travel roughly one and a half times as far through solid oak as through water. A 12.5-centimeter shot from an eighteen-pounder gun, fired with the full charge of gunpowder at 365 meters' range, could penetrate ninety centimeters (thirty-five inches) of oak, yet barely sixty centimeters (twenty-five inches) of water.

This effect was used in wooden warships, in which a passageway right around the waterline known as a "carpenters'

walk" was kept clear of stores so that shot holes could readily be plugged. Holes positioned higher up would not endanger the ship, while the hull below water was protected by the sea.

W. Roderick Stewart
Unicorn Preservation Society
HM *Frigate Unicorn*, Dundee, United Kingdom

IS ANYBODY THERE?

I have two parents, four grandparents, eight great-grandparents, and so on. If I drew a family tree going back ten generations, I would have to make space for a top line of 1,024 ancestors. At thirty generations I would expect to see a line of over a billion ancestors. If I tried to research my family back forty generations (only about a thousand years) I would be searching for the names of vastly more people than have ever lived. This is impossible, of course, but everyone has two parents, so what exactly is wrong with my reasoning?
Steve Pulsford

The simple answer is that some people marry their cousins or half-cousins. If you can have shared ancestors at the close proximity of cousin level, then imagine the number of shared ancestors there would be going back forty generations.

Although the questioner's calculation is strictly correct, it makes the mistake of assuming that all the names on the family tree are unique, when in fact there will be individuals who appear many times over. For example, if your parents were cousins you would only have six great-grandparents and not eight; and at the tenth generation point, assuming no other shared lineage, there would be 768 grandparents, not 1,024. Although people tend not to marry their cousins, you would only ever have to go back a handful of generations to find common ancestors, and the further back you go, the more repetition there would be.

Jim Rogers

Instead of starting in the present day and working back, start back in time and work forward. Two young couples are marooned on an uninhabited desert island. With plenty of food and not much to do, they have ten children each—equal numbers of boys and girls—who later pair up and have ten children each. The island population is now 124, made up of one hundred children, twenty parents and four grandparents; and not as the questioner suggested, two hundred parents and four hundred grandparents.

The reasoning is valid only if no one is related. But, as on the island, vast numbers of us share common ancestors. In fact, because the questioner has a British-sounding name, he is possibly a distant cousin of mine.

Gordon Black

One way to resolve this problem is to examine a detailed family tree. A good example is that of the United Kingdom's Prince Charles, as given in *The Lineage and Ancestry of H.R.H. Prince Charles, Prince of Wales* by Gerald Paget (Edinburgh: Charles Skilton, 1977).

His parents are third cousins because each has Queen Victoria and Prince Albert as two of their sixteen great-great-grandparents. This means that Charles has thirty and not thirty-two different great-great-great grandparents. Then it gets more complicated. Two of the great-great-grandparents

of Queen Elizabeth (Charles's mother) are also Prince Philip's great-grandparents, so the number of Charles's forebears is further reduced.

Paget attempted to identify all 262,142 individuals who comprise seventeen generations of Charles's family. He succeeded with all but 83,800. Of those he did identify he discovered not 178,342 individuals, but just 11,306. The difference (167,036) was due to duplication by intermarriage. So the number of ancestors is only 6 percent of what might have been expected. This proportionate reduction increases the further back you look. Go back far enough and we are all twigs at the bottom of the same tree.

J. R. Johnstone

People marry people they know from similar geographical, religious, or social backgrounds—so marrying cousins is reasonably common. If your parents are first cousins this instantly duplicates a quarter of your ancestral tree. Allowance must also be made because the length of a generation is variable. Oldest sons of oldest sons breed faster than youngest sons of youngest sons because they can pack more generations into a century. Add to this those cases where someone has an elderly father and a teenage mother and the generations rapidly get hopelessly mixed. The same person could be your grandfather on one branch of the family tree, and your great-great-grandfather on another.

If you tried to trace every branch back a dozen generations you would actually end up with a complex network. The smaller the population of target ancestors, the more convoluted it becomes. While most people in the past stayed in one place, a small percentage such as soldiers, sailors, and members of parliament were mobile, ensuring crossbreeding nationwide over a period of centuries.

It is more revealing to calculate the probability of a given person who was living hundreds of years ago being your ancestor. When writing for the Society of Genealogists, I built a computer model to give approximate figures for dates in the past.

In England the population fell to its lowest level for centuries at the time of the Black Death in A.D. 1348–49. My model revealed that if you have English blood in your veins, the chances are high that any named survivor of the plague is your ancestor (assuming they had children). Go back to 1066 and the model shows that everyone then living in England who had descendants is your ancestor and mine—and every other descendant (including the majority of readers of the UK edition of this book worldwide) is our cousin.

Something similar no doubt goes for all Europeans whose ancestry dates back to the time of Charlemagne (A.D. 742–814). Further back, the Roman armies in England included many from the Middle East and Africa, ensuring an even wider spread of ancestors.

Chris Reynolds

INDEX

ABOUT THE AUTHOR

New Scientist is a science magazine for everyone, young and old, amateur and professional. With a worldwide readership of more than half a million, it is among the most popular of all popular-science magazines.

Mick O'Hare is the production editor of *New Scientist* and the editor of the magazine's previous international bestsellers *Does Anything Eat Wasps?* and *Why Don't Penguins' Feet Freeze?* He lives in London.